Geophysical Development Series
Volume 2
Series Editor G. M. Hoover

Seismic Modeling of Geologic Structures

Applications to Exploration Problems

Stuart W. Fagin

PART I MODELING THEORY AND PRACTICE
PART II CASE HISTORIES

SOCIETY OF EXPLORATION GEOPHYSICISTS
P.O. BOX 702740, TULSA, OK 74170-2740

Library of Congress Cataloging in Publication Data

Fagin, Stuart William.
    Seismic modeling of geologic structures: applications to
exploration problems / Stuart W. Fagin.
        p.   cm. — (Geophysical development series; v. 2)
    Includes bibliographical references and index.
    ISBN 1-56080-050-X
    1. Seismic prospecting—Simulation methods.   I. Title.
II. Series.
    TN269.8.F34   1991
    622'.1592—dc20   91-28254
                                                                    CIP

ISBN 0-931830-41-9 (Series)
ISBN 1-56080-050-X (Volume)

Published 1991.

Printed in the United States of America

To Alice, for putting up with all this.

Table of Contents

Contributors to Case Histories
PART II

R. Bryan

F. Bussemaker

C.J. Callahan

S.W. Fagin

S.M. Greenlee

C.E. Hall

J. Johnson

S.H. Lingrey

B. McClellan

D.A. Medwedeff

P.F. Morsel

E.A. Nosal

G.W. Purnel

J.M. Reilly

K.W. Rudolph

P.M. van der Made

P. van Riel

S. Whitney

R.W. Wiley

B.E. Winkelman

D.M. Ziljstra

L.J. Zimmerman

PREFACE

Seismic interpretation apparently is becoming primarily a geologic rather than a geophysical skill. This observation has been true from the moment seismic reflection data were displayed as a continuous record with the intention of creating an image of subsurface structure. The imaging advances that have occurred in the past two decades only reinforce the tendency. More effective migration algorithms making use of faster and less expensive computers, as well as high-fold and, in particular, 3-D data all serve to make the seismic picture better. As the image increasingly reveals more geology, the geologic skills become more crucial to the task of extracting the information made available. As seismic artifacts such as multiples, sideswipe, and raypath distortion effects are successively eliminated from the image, the geophysical sophistication of the interpreter becomes increasingly less important. At first glance it would seem that these tendencies can only intensify as these technological trends continue.

And yet the depiction of complex structures remains elusive. Migration programs have been developed that can manage the severe raypath bending attendant with complex structures. Moreover, the ever decreasing costs of computation make the application of these programs increasingly more feasible. Unfortunately, to benefit from these imaging approaches requires, a priori, an increasingly more precise definition of the velocity field which often is, in itself, an expression of geologic structure. Therefore, before we can create the image, we require an understanding of what the image is supposed to show. This circumstance implies that the preparation of the seismic image has become, and will likely remain, inextricably bound up with its interpretation.

Similarly, although the development of 3-D imaging has largely eliminated the crucial problem of sideswipe, 3-D is still primarily restricted to production and development applications. Though 3-D surveys are increasingly used before drilling the exploration well, for some time to come exploration prospects, and certainly leads, likely will have to be defined and defended using 2-D seismic grids. For the time being, sideswipe is here to stay.

As a result of these and other problems, the seismic depiction of complex structures leaves much to be desired. At times only minor portions of structures can be discerned and in many cases reflections are downright misleading. However, despite these failings, every reflection on each seismic line represents a sampling of subsurface structure and therefore has the potential for constraining structure in some fashion. With this outlook in mind, the technology presented in this book, seismic modeling, has a method of operation which runs directly counter to the trends just described. Instead of looking upon the seismic section as a picture of structure, each reflection is traced back to the surface it originated from and, in so doing, the structure is defined.

For several years at Exxon Production Research Company I have applied this technology to complex structure problems. Although much literature was available on modeling theory, I quickly discovered that there was little documentation on the subject of how, when, where, and why to model. Appropriate case history reports were difficult to find even within company files. Therefore, the goal of this volume is to bring together a series of case histories which, in their variety, illustrate how modeling can be used to better define structures. To achieve this goal the competitive nature of petroleum exploration, which often prevents the release of such examples, had to be overcome. Fortunately, over the past year enough examples have been collected from contributing authors to make this volume possible.

In addition, to provide a thorough context for these case histories, a review of modeling theory and practice is presented which comprises the first part of the volume. In his recent best seller on theoretical physics, Stephen Hawking observes that the potential readership of any text is reduced tenfold with each additional equation included. In this spirit, theoretical material in the first part of the book is, with some exceptions, presented in an informal and descriptive manner. Because the intended audience for the volume extends to the geologist/interpreter with limited background in seismic theory, this sort of treatment should be appropriate.

ACKNOWLEDGMENTS

I am deeply indebted to many individuals for the ideas and work presented in this volume. My interest in seismic interpretation was initiated by the enthusiasm and teaching skill of Milo Backus at the University of Texas. I have also benefited enormously from interaction with the geophysical staff at Exxon Production Research Company. P. M. Shah introduced me to many modeling issues. J. R. Berryhill brought to my attention the relationship between modeling and wave equation migration. E. W. Peikert and Z. J. Nikolic made me aware of the strengths and limitations associated with seismic velocity interpretation. Particularly beneficial has been my association with G. H. Weisser and H. Yorston both of whom have extensive experience in the seismic interpretation of structures.

The ideas and viewpoints presented in this volume developed primarily as an outgrowth of the projects I have engaged in and the people I have worked with. At Exxon I have been fortunate to be able to collaborate with individuals who were always enthusiastic about exploring new ideas and techniques that aid in the definition of structure. In this regard, I owe much to J. A. Dickinson, V. Khare, J. R. Krebs, and K. J. Young. I have also greatly benefited from discussions with M. S. Ephron, R. Gonzalez, and Y. C. Kim. I also wish to acknowledge the work of present and former members of the Sierra Geophysics Corporation, particularly D. Hadley, S. Maher, and G. Lundquist, whose early development of a usable 3-D ray-tracing system has had an enormous effect in promoting modeling throughout the petroleum industry.

Perhaps most importantly, I wish to thank the twenty-two case history authors. Without these examples the theme of the book, that seismic modeling is a useful procedure for defining structures, would have been unsupported.

This volume was reviewed by G. M. Hoover, R. Gonzalez, B. A. Hardage, V. Khare, N. Marshall, M. F. McGroder, A. S. Meltzer, P. M. Shah, T. J. Stark, G. H. Weisser, and H. Yorston. The final product has been greatly improved because of their efforts. C. Harrell and M. Kirby were a considerable help in the preparation of illustrations and I am deeply grateful to Lynn Griffin for her extraordinary effort in preparing the manuscript for publication. Finally, I am grateful to Exxon Production Research Company, Exxon USA, and Esso Exploration and Production U.K. for permitting release of data, and to Exxon Production Research Company, for the time and support provided to complete this project.

PART I

Modeling Theory and Practice

CHAPTER 1

The Need for Seismic Modeling of Geologic Structures

THE PHOTOGRAPHIC IMAGE

Seismic reflection profiling is unique among the geophysical methods in what it proposes to achieve: a direct cross-sectional image of the layered portions of the earth's crust. A measure of the success that has been achieved toward this goal is that a geologist, with little understanding of the imaging method, can interpret the data in a way that uses his geologic experience to solve problems. Although we take this adaptability for granted, there is a great contrast with the data treatment of other geophysical methods used in petroleum exploration; primarily gravity and magnetics. In these methods detailed interpretation requires modeling and inversion approaches which in turn require greater geophysical sophistication on the part of the interpreter.

For example, consider the section shown in Figure 1 from the East Texas salt basin. Any geologist, given some notion of scale, will recognize the presence of a piercing salt diapir in the middle of the section. If the geologist is familiar with structural patterns associated with diapirs, he will recognize flanking withdrawal synclines which migrate toward the dome with time and signify an inwardly collapsing salt reservoir. In addition, the interpreter will observe a low-relief salt pillow on the right which does not pierce the overlying section, and a moderate-relief structure to the left in which piercement level is more difficult to judge.

All of these observations can be made by simply accepting the section as a picture of subsurface structure in vertical cross-section. In other words, the section is regarded as if it were a photograph of a vertical cliff face running along the line of the seismic section. In the discussions that follow, this view of the seismic section is referred to as the "photographic image".

In summary, without any knowledge of the imaging method, a well-informed geologist can learn much by viewing the seismic section as a photographic image. This statement does not suggest that the interpreter is better off without such knowledge only that seismic imaging has progressed so far that much can be discerned without it. Because of this success, accepting the section as a photographic image is properly the method of first resort of the interpreter. The burden of proof rests on the side of those who would contend that a particular seismic event is not properly imaging structure. But there are limits to the detail and accuracy of an interpretation made solely by treating the seismic section in this way. These limits are made apparent when we take a closer look at the figures referenced here.

In Figure 1 the base-of-salt reflection shows several apparent basement structures, each of which is localized beneath a salt structure. Most prominent is the "basement anticline" below the diapir in the center of the section. This basement anticline is a velocity pull-up which arises because salt velocities are greater than

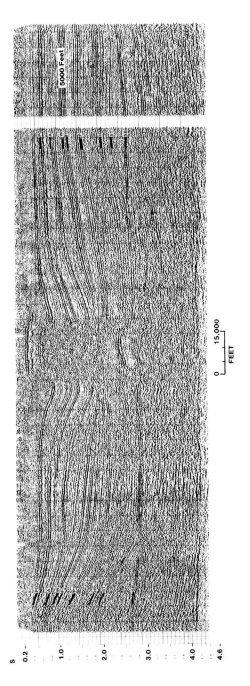

Fig. 1. Migrated seismic section depicting several salt structures.

those of the surrounding section. Traveltimes to the base of salt are therefore shorter.

Velocity anomalies can be used to infer something about overlying structure. The pull-up below the structure on the left is clearly of lower magnitude than in the central diapir. We may infer that the seismic wavefront traveled through a smaller thickness of salt and that top of salt is therefore at a much lower level than the central dome. In contrast, the low-relief structure on the right side of the section, surprisingly, depicts a velocity sag rather than pull-up (we're assuming that the depth structure of the base-of-salt is, in reality, planar). How can this be explained? Note that the salt structure is of very low relief and the salt "replaces" only narrow stratigraphic intervals in the middle part of the stratigraphic section (The missing beds are truncated by onlap in these intervals). A reasonable explanation for the sag is that the interval velocity of this replaced section is actually faster than that of the salt. (This section is presented, and these distortions are pointed out in Tucker and Yorston, 1973).

Next, consider the section shown in Figure 17 (Case History 2, page 108) which was taken over a fold and thrust belt. Noted on the section four structures which might be prospects. However, only two of these structures are genuine while the other two are a result of velocity effects. In their case history presented in Part II of this volume, McClellan et al. relate how this distinction is made.

Finally, consider the sections in Figure 13a (Case History 9, page 232) and in Figure 64 (Chapter 4, page 74). Figure 13a depicts a deepwater salt sill with prominent top-of-salt and base-of-salt reflections. Below the base-of-salt is a reflection dipping steeply to the right. If the section is viewed as a photographic image, then the subsalt section would be interpreted as rotated, truncated, and highly disconformable with the base of salt. In fact, this reflection below the base-of-salt is a second subsalt reflection recorded from out of the plane of section, or in other words, a sideswiped reflection. (Sideswipe reflections are discussed in detail in Chapter 4.) Proof of this is shown in Case History 9 presented in Part II.

Similarly Figure 64 (Chapter 4) shows two prominent reflections in the middle of the section dipping to the left and adjacent to a fault-bounded uplift on the right. These reflections are known to be associated with the same reflector and have been interpreted as evidence of repetition resulting from tectonic compression. In fact, as is shown in Chapter 4, both of these reflections are sideswiped from block edges located in and out of the cross section, and the horizon is not structurally repeated.

These examples demonstrate that there is a point at which the seismic image departs from a photographic image. In order to obtain structural information beyond this point, the manner in which the structure was imaged must be considered.

THE IMAGING MARGIN

The degree that a section departs from a photographic image depends on the complexity of the structure to be imaged and the effort applied to the imaging. Structures which (a) incorporate units with highly contrasting velocities, (b) have steep dips, (c) are composed of closely spaced folds and faults, and (d) do not have well-defined strike directions all cause imaging problems. The section departs from a photographic image because, the ability of seismic reflection imaging to handle these problems is limited. These limits can be broadened, but never eliminated, with the application of higher-effort imaging approaches such as 3-D seismic, depth migration, or prestack migration.

The term "complex structure" is used here for those structures which cannot be easily imaged because of the structural characteristics described. Petroleum exploration occurs in many settings which include complex structures. The most common examples are mobilized salt provinces, fold and thrust belts, and extensional glide-plane provinces.

In each of these tectonic settings, exploration advances from the highly visible to the slightly visible targets. The development of new prospects depends on the explorationist's ability to see or infer something previously unrecognized. For these reasons, exploration in complex structure areas soon concentrates along an "imaging margin" along which the seismic evidence for the prospect is slight. In areas of salt mobilization, the imaging margin (and exploration) progresses from depicting targets along the crest of nonpiercing structures, to increasingly more steeply dipping targets along the flanks of diapirs, to targets beneath salt overhangs. Current and future efforts are likely to concentrate on targets completely covered by salt sills. Similarly, in thrust-belt exploration the imaging margin advances from surface anticlines to increasingly more complex and deeper imbricates.

THE ROLE OF SEISMIC MODELING

As exploration concentrates along the imaging margin, the need develops for an analysis which derives structure without relying on the seismic section as a photographic image. This need is satisfied by the role of seismic modeling. The approach used, shown diagramatically in Figure 2, is to treat the seismic section as a display of recorded reflections rather than as a photographic image. By predicting the reflections that should develop from a particular depth model and

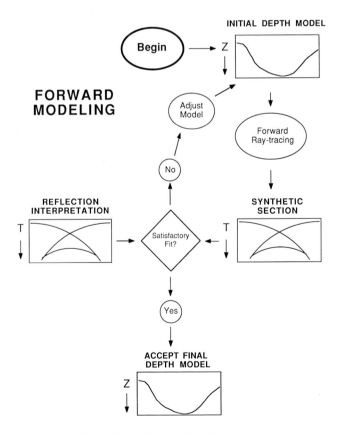

Fig. 2. Seismic modeling diagrammed.

comparing these reflections with the appropriate real data the interpreter is able to determine whether the depth model is compatible with the data. If not, the model is modified and, by successive iterations, a structure derived where the seismic response closely recreates these reflections.

In a sense this approach is a throwback to the way seismic reflection data were treated at their inception and even into the postwar era. Prior to the advent of digital recording and CDP acquisition, data displayed along analogue tapes were visually examined to estimate arrival times to key reflections to determine gross structure. Reflections were migrated to their proper lateral position by graphic methods (see the representative papers in Gardner, 1985 for examples). Traces were generally not composited to create a subsurface image. In a similar way, modeling utilizes an approach which directs the interpreter to examine the section as a composite of reflections rather than as an image of geologic structure.

The development of digitally recorded data and the migration algorithms to operate on them made these techniques largely obsolete, and the modeling of most structures unnecessary. For example, it would not be a productive use of the interpreter's time to model a bow-tie structure observed on an unmigrated section, in order to derive a syncline that is likely to be accurately imaged on the migrated section.

When should the interpreter suspect inaccurate imaging of a structure? The two most important limits for seismic migration programs are their inability to image sideswipe reflections (if the data are 2-D) and their inability to image reflections with complex raypaths. If either of these problems hinder satisfactory definition of a prospect, the use of seismic modeling should be considered.

Seismic modeling is also used to design acquisition programs which will ensure the recording of reflections from key portions of structures. In particular, the effect of varying line, spread, and array lengths can be evaluated for a specific structure. In a similar manner, modeling is being increasingly used to support the processing of seismic data over complex-structure areas. Just as the effectiveness of interpretation is limited without knowledge of the imaging process employed, so the ability to effectively process is greatly limited without bringing to bear on the attempt what is known about the structure. The model is the vehicle by which this is done and the approach can be thought of as "model-guided processing". May and Covey (1983) presented an example of this approach as part of an effort to image a salt dome flank. Canal and Diet (1987) present an example of how modeling is used to establish and portray a velocity field for 3-D depth migration. In these cases the model was used to define processing parameters, such as stacking and migration velocities, and as a monitor of processing performance.

The goal of this book is to demonstrate, by example, how this technology can be used to define hydrocarbon traps in complex structure areas. The book is almost entirely devoted to ray-trace modeling, because that is the most appropriate modeling form for structural definition. Part I reviews the basic methods of modeling. Chapter 2 describes the varieties of ray-tracing approaches and the circumstances under which each is applied. Chapter 3 reviews major issues the interpreter faces in building a model. Chapter 4 discusses the interpretation, mapping, and migration of reflection structure. Chapter 5 reviews the limitations, errors, and pitfalls of modeling.

The use of modeling can only be effectively demonstrated by convincing examples. Seismic structural modeling studies often are presented in association with schematic structures to demonstrate their seismic characteristics [Hron and May (1978) and Withjack et al. (1987)]. In addition, modeling is commonly used to create synthetic data sets which in turn are used to investigate the effects of data processing techniques. [See Taner et al. (1970) and Miller (1974) for early examples of what is now a quite routine application of seismic modeling]. Although these applications can be quite useful, the need still exists to show how modeling can be used to define specific structures. Skeen and Ray (1983) and Yancey and McClellan (1983) are two such examples, but there is a strong need for many more. Part II contains ten case histories from contributing authors, most of whom are directly involved in petroleum exploration. The purpose of these case histories is to demonstrate the variety of ways in which modeling can be employed to define structure.

CHAPTER 2

Seismic Modeling Approaches

A variety of modeling approaches exist which attempt to simulate the seismic response to subsurface structure. The approaches are distinguished by the assumptions they make about subsurface structure and the seismic data to be modeled. To be effective, the interpreter must have a clear understanding of which approach is appropriate for the problem at hand. In this chapter modeling methods are reviewed and contrasted.

Because ray tracing is the modeling method most applicable to structural problems almost all the discussion in this section, and throughout the book, is devoted to it. Wave theory modeling is briefly reviewed and contrasted with ray tracing at the end of this section. In addition, a review and example of physical modeling is presented by L. Zimmerman in Case History 7.

The different ray-tracing modes are diagrammed in Figures 3, 4, and 5 and Table 1. These figures diagram the ways in which ray-tracing modes differ. Figure 3, a flow diagram, shows that specific ray-tracing modes are associated with the particular data types that result from the seismic processing stream. Shot (sometimes called "line ordered") gathers and CMP gathers are modeled by variable incidence rays where the raypaths reflect off model surfaces at varying angles. The stacked section is represented by normal incidence rays which reflect only at right angles to model horizons. The time-migrated section is represented by the image ray which is normal to the ground surface but bends across model horizons. The depth-migrated section is represented by vertical incidence rays which extend, without bending, from the ground surface to model horizons. These vertical incidence rays are not defined by ray tracing but rather represent the depth-to-time and time-to-depth conversions that result when depth or time maps are divided or multiplied by velocity maps. All of these concepts are elaborated upon in this chapter.

Figures 4 and 5 organize the ray-tracing modes, and the names applied to them, according to whether they simulate offset (variable incidence rays) or zero-offset (normal incidence rays or image rays) conditions, whether they work in a forward or inverse manner, and whether they operate on a two- or three-dimensional model. The terms used in these diagrams will be referred to in the rest of the chapter.

Table 1 calls attention to the assumptions attendant with each ray-tracing mode. Fewer assumptions are required for higher-effort modeling modes. The three assumptions which can underlie a modeling mode are: (1) the CMP stacked section is equivalent to zero-offset, normal incidence acquisition, (2) reflection raypaths are in the plane of the seismic section (they are not sideswipe reflections), and (3) reflection arrival times and lateral position are not subject to image ray effects. (Image rays, which are discussed in Section 2, are subject to additional assumptions.) The rigor of ray tracing modes is greatest with offset ray tracing in three dimensions, (e.g., 3-D tomography) where none of these assumptions are made. Rigor is least in performing a vertical incidence time/depth conversion where all these assumptions are made.

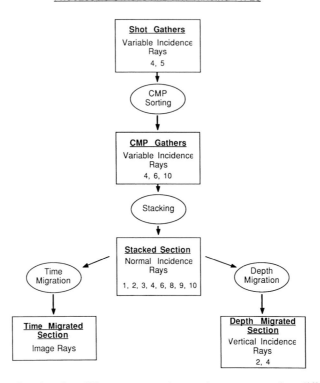

PROCESSING STAGES AND RAY-TRACING TYPES

Fig. 3. Diagram showing that different ray-tracing modes correspond to different stages in the processing stream. Numbers indicate the case histories in which the different ray-tracing modes are represented.

ZERO OFFSET MODELING

	FORWARD	INVERSE
2-D	2-D FORWARD MODELING 1,2,3,4,5,6,8	HORIZON MIGRATION
3-D	3-D FORWARD MODELING 9,10	MAP MIGRATION 4,8,9,10

Fig. 4. Diagram categorizing zero-offset ray-tracing modes. Map migration is the name applied to 3-D, zero-offset inversion. Numbers indicate the case histories in which the different ray-tracing modes are represented.

RAYPATH FUNDAMENTALS

The seismic raypath

The manner in which a seismic wave propagates through the subsurface is predicted by Huygen's Principle, which states that each point on the wavefront acts as the source of an entirely new wavefront. Figure 6 shows a wavefront, propagating through a constant velocity media, 2 seconds after a source is energized. (The wavefront is circular in cross-section, but is spherical in three dimensions). To predict the position of the wavefront at 3 seconds, each point on the 2 seconds wavefront is considered a new source. Circles centered on these points with radii corresponding to 1 second of traveltime represent the wavefront associated with each new source. The large envelope tangential to these smaller wavefronts represents the advanced position of the seismic wave at 3 seconds. The portions of the smaller wavefronts not coincident with this envelope are out of phase and cancel one another leaving only this single wavefront.

Having drawn a succession of wavefronts, we can also draw their associated normals. These normals will lead back to the source location because the normal is the radius that connects a wavefront with the "source point" on the preceding wavefront. These normals which connect temporally successive wavefronts are

OFFSET MODELING

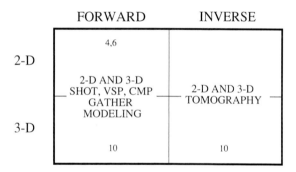

Fig. 5. Diagram categorizing offset ray-tracing modes. Tomography is the inversion of offset arrival-time information. Numbers indicate the case histories in which the different ray-tracing modes are represented.

Table 1. Assumptions associated with different ray-tracing modes.

RAY TRACING MODES

DATA TYPE	RAY TRACING TYPE WHICH SIMULATES	ASSUMES CMP STACK = NORMAL INCIDENCE	ASSUMES NO SIDESWIPE 2-D	ASSUMES NO SIDESWIPE 3-D	ASSUMES NO IMAGE RAY EFFECT
Shot, VSP, and CMP Gathers	Variable Incidence Ray	No	Yes	No	No
Stacked, Unmigrated Section	Normal Incidence Ray	Yes	Yes	No	No
Time Migrated Section	Image Ray	Yes	Yes	Yes	No
Depth Migrated Section	Vertical Incidence Ray	Yes	Yes	Yes	Yes

termed raypaths. The raypaths for the wavefront in Figure 6 are shown in Figure 7.

Wavefronts are real in that they describe those portions of the seismic disturbance that are contiguous and in phase. They are visible to the eye where a disturbance causes a change in optical reflectivity of the propagating media. Such is the case where we detect ripples in a pond because light reflects differently where the wave crests stand in relief. Raypaths, on the other hand, are conceptual devices which allow us to understand, describe, and illustrate wave propagation. When showing how a seismic wave propagates raypaths are easier to draw than wavefronts. However, a single raypath only describes the propagation of the portion of the seismic disturbance which leads from one particular source to one particular receiver.

Raypath trajectories

Three rules govern the trajectory of raypaths:

(1) Raypaths are unbent in a constant velocity medium.
(2) Raypaths bend in accordance with Snell's Law as they cross velocity interfaces.
(3) Raypaths reflect at an angle equal to the incidence angle when they encounter impedance interfaces.

HUYGEN'S PRINCIPLE

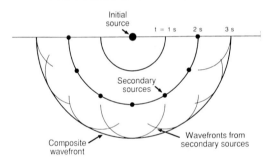

Fig. 6. Schematic describing Huygen's principle. Huygen's principle states that each point on a wavefront acts as a source for a new wavefront. The dots along the $t = 2$ s wavefront are the sources for the seismic waves whose wavefronts 1 s later ($t = 3$ s) are shown by the smaller arcs. These wavefronts interfere to form the continuous enveloping wavefront at $t = 3$ seconds.

RAYPATHS AND WAVEFRONTS

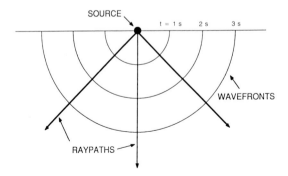

Fig. 7. Schematic showing the relationship between raypaths and wavefronts. Raypaths are normal to wavefronts and lead back to the seismic wave source.

These three rules are explored in detail.

Rule 1.—Raypaths are unbent in a constant velocity medium. Figure 7 illustrates wave propagation through constant velocity media with concentric wavefronts. Because the wavefronts are concentric, the raypaths are straight and lead back to the source located at the center of the circular wavefronts.

Rule 2.—Raypaths bend in accordance with Snell's Law as they cross velocity interfaces. The nature of this bending is diagrammed in Figure 8. In the figure, a series of wavefronts are shown where they impinge upon an interface between media of different velocities. The wavefronts are shown as planar but can be thought of as local portions of wavefronts which are regionally spherical.

The spacing of successive wavefronts is greater in the underlying high-velocity media signifying greater distance traveled for the time increment, Δt. In addition, the wavefronts, and their associated raypath, bend, or refract, when they encounter the higher velocity media. The reason for the wavefront bending is shown by the local wavefront arc representing a traveltime of delta t from a point source located at C on wavefront $T + 2\ (\Delta t)$. The new wavefront at $T + 3\ (\Delta t)$ is formed by joining a tangent to the arc at point F, with point D where wavefront $T + 3\ (\Delta t)$ encounters the velocity interface. This results in a more vertical orientation of the wavefront and a more lateral propagation in the high velocity media.

The amount of raypath bending depends on the velocity contrast across the interface and is predicted by Snell's Law:

$$\frac{\sin (a1)}{V1} = \frac{\sin (a2)}{V2}$$

where $a1$ equals the angle of incidence and $a2$ equals the angle of transmission. These are the angles between the incoming and transmitted raypaths, respectively, and the normal to the interface. Snell's Law is derived in Figure 8.

When the angle of incidence, $a1$ in Figure 8, reaches a critical value, such that the angle of transmission, $a2$ in Figure 8, is 90 degrees, there is no transmission of the wave into the lower media. Substituting 90 degrees for $a2$ in the above equation, we derive this critical angle of incidence as:

DERIVATION OF SNELL'S LAW

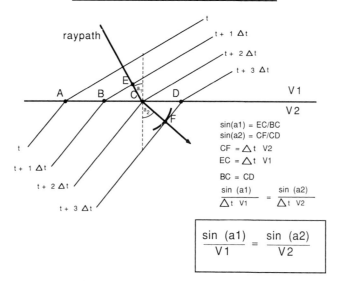

Fig. 8. Derivation of Snell's Law.

$$\sin a \ \text{(critical)} = V1/V2.$$

When the angle of incidence is critical, raypaths arise which travel along the interface (at the velocity of the underlying media) and reenter the upper, low-velocity media along a continuum of points along the interface (see Figure 9). The seismic wave represented by these raypaths is termed a refracted wave or head wave.

From Snell's Law we note that raypaths bend toward the interface normal as they enter low-velocity media, and away from the normal as they enter high-velocity material. Because the most common subsurface condition is increasing velocity with depth, raypaths tend to bend away from the normal as they propagate into the earth and toward the normal as they return upward to the surface (see Figure 10). Also, if a raypath is normal to an interface, the angle of incidence is zero. Then $\sin(a1)$ in Snell's Law is zero and $\sin(a2)$ and $a2$, the transmission angle, must also be zero. In other words, when the raypath is normal to an interface there is no raypath bending in the transmitted wave.

Raypath bending results in increased raypath length in the higher velocity layer (as compared to an unbent ray) with a consequent shortening of traveltime. In fact, the raypath predicted by Snell's Law between any two points, can be shown to be the one with the least traveltime between those two points. This is known as Fermat's principle and is employed in some ray tracing procedures (e.g., Waltham, 1988). Raypath bending, which greatly complicates complex structure imaging, is most severe when the seismic wave transmits through steep interfaces separating units of widely contrasting velocities. One common example occurs in attempting to see below the steep flanks of a salt dome as shown in Figure 11.

Rule 3.—Raypaths reflect at an angle equal to the incidence angle when they encounter impedance interfaces. The impedance of a medium is the product of the

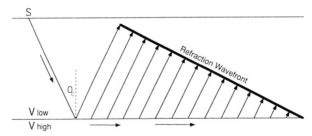

Fig. 9. Raypaths associated with refraction arrivals. Raypaths shown contain three segments. The first segment initiates at source S and travels at velocity V(low) to the velocity interface. The second segment travels along the velocity interface at velocity V(high) and the third segment returns to the ground surface at velocity V(low). The raypath enters and emerges from the interface at the critical angle Oc.

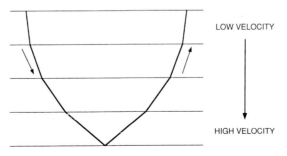

Fig. 10. Reflection raypath showing the tendency for raypaths to be more vertical as they approach the ground surface. It may be inferred from this diagram that a media with continuously increasing velocities would result in a curved raypath.

mediums velocity and density. When a seismic wave encounters an interface between units of different impedance, part of the seismic energy is directed back toward the surface in the form of a reflected wave. Figure 12 shows that the incidence angle O_i equals the reflection angle O_r.

These three raypath rules govern wavefront propagation. Note that each of the rules is subject to the principle of reciprocity: the raypath remains unchanged regardless of travel direction. In a seismic experiment, if the position of the source and receiver are switched, the raypaths and traveltimes remain unchanged.

Tracing the ray

The ray tracing problem is stated as follows: given a model of subsurface structure and velocities, a source location, and a receiver location, find the raypath that reflects from a particular model surface and leads from the source to

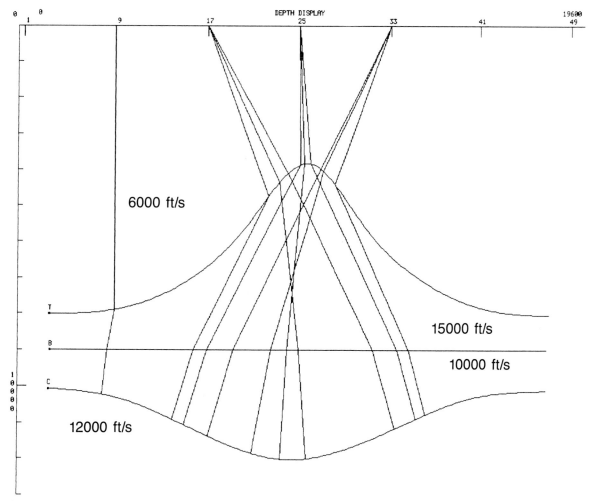

Fig. 11. Raypaths demonstrating strong raypath bending with transmission through the flank of a salt dome. The model contains three surfaces; a domal top-of-salt, a flat base-of-salt, and a synclinal subsalt reflector. The section above salt has a velocity of 6000 ft/s, the salt 15 000 ft/s, and the subsalt section a velocity of 10 000 ft/s. Normal incidence raypaths from the subsalt reflector are shown at stations 9, 17, 25, and 33. (Raypaths that do not look quite normal are the result of vertical exaggeration in the diagram.) Note that stations 17, 25, and 33 each record three raypaths from the subsalt reflector, indicating the reflection is multivalued in time. (This is an example of a bow-tie reflection which is discussed in depth in Chapter 4). Note the extreme raypath bending exhibited by the leftmost ray from station 17 and the rightmost ray from station 33 where they refract across the top-of-salt. Both of these rays encounter the top-of-salt at large incidence angles. In contrast rays which encounter the top-of-salt at smaller incidence angles (rightmost ray of station 17) refract much less. To properly image subsalt structure by wave equation migration the velocity field (and hence the top-of-salt structure) would have to be accurately estimated.

the receiver. The three raypath rules determine how the raypath will bend and reflect through the subsurface. Despite these constraints, to derive the raypath directly given a generalized subsurface structure is difficult. An alternative strategy is generally adopted: the raypath is determined by trial and error, or iteratively, until a satisfactory path is found.

The offset case.—Consider the structure in Figure 13 which consists of a right-dipping surface A located above left-dipping surface B. Both of these surfaces bound media of constant velocity. We wish to determine the raypath for a reflection from surface B that would extend from the source position S to the receiver position R. The ray is initiated at the source location (although considering reciprocity the ray could be initiated at the receiver location). An initial takeoff angle is assumed, perhaps vertical to the ground surface. The ray is shown as ray number 1. The ray is extended vertically downward until refracting surface A is encountered. Snell's Law predicts the raypath bending across surface A. The raypath maintains this direction until reflecting surface B is encountered at reflection point C thus completing the downgoing ray.

The upgoing ray initiates at this reflection point and departs at a reflection angle equal to the incidence angle. This direction is maintained until the upgoing ray encounters surface A where the ray again refracts in accordance with Snell's Law. This final direction is maintained until the upgoing wave encounters the ground surface at point D.

Point D is a great distance to the left of receiver location R. In order to get closer, and obtain a better estimate of the raypath, the departure angle from the source must be altered. Given the location of point D, the departing raypath should be rotated toward the receiver position as in raypath 2. By the same

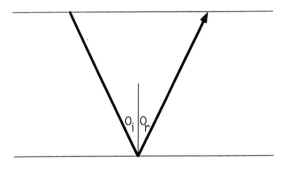

Fig. 12. Reflection raypath showing that the angle of incidence (O_i) equals the reflection angle (O_r).

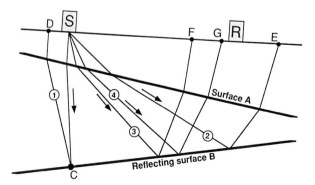

Fig. 13. Diagram showing how a ray reflecting off of surface B may be traced from source S to receiver R. Several departure angles from S are attempted until one lands acceptably close to the target (at point G).

procedure as described for raypath 1, the downgoing and upgoing segments of raypath 2 are traced as well as the emergence point E. Point E is located to the right of R so the departure angle was rotated too much. Two more ray-tracing iterations are shown with the last, raypath 4, emerging very close to R at point G. In practice, if the emerging ray is within a specified "capture radius" the ray is considered "captured" (in other words, close enough to be considered the ray recorded at the receiver).

Having captured a ray, the program then stores the rays parameters; the coordinates of the raypath endpoints, bendpoints and reflection points, and, most importantly, the raypath traveltime. In addition, an estimate may be made of reflection amplitude, given the reflection coefficient of the reflecting surface and length of the raypath.

In Figure 13, the general condition of tracing a ray from a shotpoint to a receiver from any offset was considered. However, when modeling a stacked section, we wish to simulate zero-offset acquisition. Although the raypath could be determined in the same way as was done in the offset case, the ray tracing can be made simpler.

The zero-offset case.—Figure 14 shows a source and receiver located at the same position. Because of reciprocity, the raypath leading from the source/receiver position to the subsurface reflection point, must be identical to that leading from the reflection point to the source/receiver. Therefore, only the upgoing or downgoing raypath needs to be traced, not both. Moreover, because the raypaths are identical and the incidence angle and reflection angle must therefore be equal, the raypath must be normal to the reflecting surface. This is why zero-offset ray tracing is often referred to as normal-incidence ray tracing (although we will see later that zero-offset "image rays" are not generally normally incident).

In normal-incidence ray tracing we, therefore, seek to find a raypath which extends from the source/receiver location to some reflection point along a specified reflector. Because the ray is known to be normal to the reflector, the ray can be initiated at the reflector and traced upward. In Figure 15 a ray is sought which would link the indicated source/receiver location with some reflection point lying along the left-dipping surface. A ray is initiated, perhaps at B, lying on the reflector. After accounting for raypath bending, the ray emerges at C, some distance from the source/receiver. The takeoff point (really a reflection point) is changed in a way that makes the emergence point closer to the source/receiver. This is done for the second iteration raypath which starts at D and emerges at E.

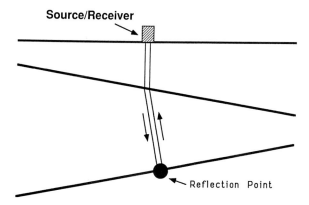

Fig. 14. Diagram showing that the upgoing and downgoing raypath are identical for the zero-offset ray. For this reason the ray must be normal to the reflecting surface or the angle of incidence would not equal the reflection angle.

If E is within the capture radius of the source/receiver, we have successfully associated a reflection point with the source/receiver location.

Because we calculate only the upgoing wave, the source seems to be located on the reflector rather than the ground surface. Indeed, the unmigrated section is often viewed as though the section was created by a continuous series of sources located along each reflector and set off at the same instant in time. (The only caveat is that arrival times are doubled on the unmigrated section because of two-way travel). This is termed the "exploding reflector" model, and is an important concept in the development of wave equation migration procedures.

Three dimensional ray-tracing considerations.—Up to this point, ray-tracing discussion has been restricted to 2-D models. Such models would be valid only if the data contained no reflection from out of the plane of the seismic section. The 2-D ray-tracing procedures are readily generalized to 3-D (Shah, 1973). However, for computational efficiency an extra step may be taken—the generation of working rays.

Figure 16 shows a 3-D model consisting of a domal subsurface reflector and the ground surface. Zero-offset rays from the reflecting surface depart at regularly spaced intervals. Note that these rays are not iteratively adjusted to converge on a particular point on the ground surface. Instead, the coordinates of the raypath are stored on file so that when a shot/receiver location is specified for ray tracing, this file is used to target an area on the reflector to begin the search for reflection points. This raypath file is termed a working ray set. Figure 17 shows a working ray set for offset ray-tracing. In this case the working rays are associated with a particular shot location as well as a particular reflector.

RAY-TRACING MODES

As previously noted, ray-tracing modes vary depending on the acquisition mode or the processing milestone of the data the modeler intends to simulate. These two issues, acquisition mode and processing milestone, are closely related as shown in Figure 3 and Table 1.

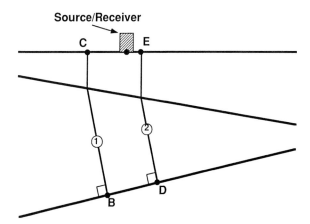

Fig. 15. Diagram showing how a zero-offset ray is traced from the lower reflecting surface to a source/receiver. A ray is traced upward from the reflector starting at point B. The ray emerges at point C too far to the left of the source/receiver. A second initiation point is tried at point D which results in an emergence point E, acceptably close to the source/receiver.

Offset ray tracing

In the description of the three acquisition modes that follows, the source is offset some distance from the receiver. The object of the ray-tracing is to find some raypath between the two.

Shot gather ray tracing.—Shot gathers, sometimes called line-ordered, or common-shot gathers, are the collection of traces associated with different geophones but the same source. This is the way the data are collected in the field.

Raypaths for a shot gather are shown in Figure 18. They are determined by searching for a raypath that leads from the source to a subsurface reflection point and to a target receiver. If the shots are thought of as receivers, and the receivers as shots, the resulting gather would be a common-receiver gather.

Shot gather modeling is used to investigate data processing steps which operate on common shot and receiver gathers, such as dip filtering and some prestack migration algorithms. Also shot gather modeling may be used to simulate array effects by positioning each receiver as if the receiver were a separate element in a geophone group.

VSP gather ray tracing.—VSP gathers are acquired by shooting from a ground surface location into geophones located in a borehole. The source may be located near the wellhead or may be offset some distance (offset VSP). VSPs are particularly well suited to analyzing complex-structure problems because their

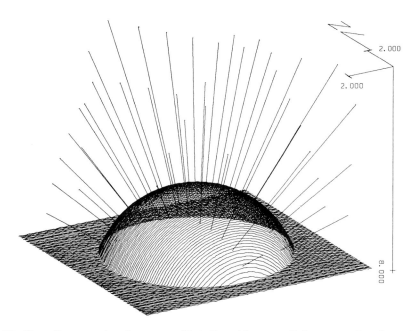

Fig. 16. Ray diagram showing zero-offset "working rays" from a subsurface dome. Working rays differ from those previously discussed in that they are not traced with the intent of achieving capture at a particular source/receiver point. Instead they are initiated along a reflector at regularly spaced intervals (for example, on 500 ft centers) with the intent of defining an emergence point on the ground surface. A file is, therefore, created in which subsurface reflection points are associated with ground-surface recording points. When ray capture to a particular source/receiver is subsequently attempted, this file is examined to identify the portion (or portions) of the subsurface where the appropriate reflection point is likely to reside. This diagram is a 3-D perspective plot in which the subsurface dome is outlined in green and the ground surface is indicated by the red rectangle. The use of such displays is discussed in Section 2.6.

configuration more readily records reflections from steep surfaces than surface seismic surveys.

Raypaths for a VSP gather are shown in Figure 19. Ray-trace modeling is commonly used in VSP surveys acquired to investigate structure. Ray trace modeling is needed because VSP surveys do not generally result in images of the subsurface structure but are instead unimaged records of reflections from surfaces whose structure is under investigation. The structure can often be best defined through modeling by deriving a solution which recreates the VSP survey.

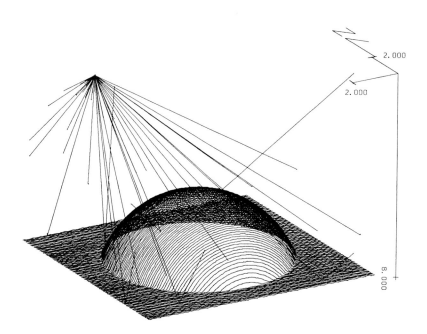

Fig. 17. Ray diagram showing offset working rays associated with a reflection from the dome and initiating from the shotpoint located along the western margin of the model area.

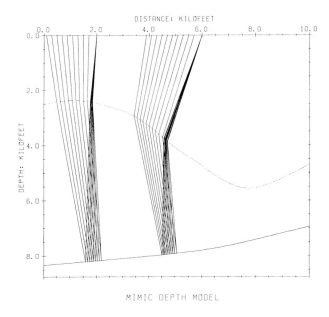

Fig. 18. Ray diagram showing a shot gather for two shotpoints located at coordinates 2.0 and 6.0. The rays are traced for a 2000 ft spread length.

Furthermore, ray-trace modeling can effectively be used to plan survey parameters such as the geophone levels in the well and the placement of the source. These parameters should be selected to maximize reflection time differences from alternative structural solutions. Indeed, modeling can tell the interpreter whether the VSP surveys can answer the questions he is interested in. Hardage (1983) makes the point clearly when he states "If VSP data are recorded in a well that penetrates complicated subsurface structure and stratigraphy, an analysis of the data should not be considered complete until a 3-D ray-trace modeling study is made."

Case History 5, by Nosal and others, is an example of the use of modeling to support the planning and interpretation of a salt proximity survey. This type of survey is a variation of the VSP where the transmitted arrivals are modeled instead of the reflected ones. The salt proximity survey is, as the name implies, well suited for defining the flanks of salt structures.

CMP gather ray tracing.—Common midpoint (CMP) gathers are simulated by CMP ray tracing. These gathers are generated by sorting shot gathers so that traces have a CMP between source and receiver. These gathers are ultimately summed in the stacking process to form a simulated zero-offset trace. Rays associated with a CMP gather are shown in Figure 20.

CMP gather modeling is used to investigate the effect of data processing steps which are performed after CMP sorting; primarily velocity analysis and stacking. For structural applications a common use is to model moveout over complex structures. Figure 20 shows that although each source/receiver pair has the same midpoint, they have different reflection points. This produces "depth point smear", and violates the first modeling assumption; that the CMP stacked section can be represented by zero-offset normal incidence raypaths.

Zero-offset ray tracing

In the acquisition modes described there is no offset between the source and receiver.

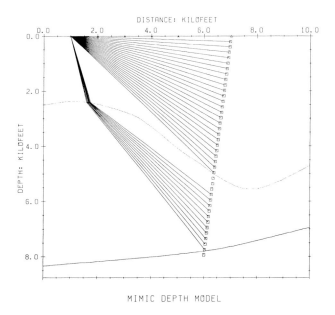

Fig. 19. Ray diagram showing an offset VSP gather. The rays extend from the source located at 1.0 to geophones located in a deviated well. Note that these rays represent direct arrivals (no reflections) from source to receiver. Reflected arrivals could be traced as well.

Normal-incidence ray tracing.—The CMP stacked section is simulated by normal-incidence ray tracing. Because normal-incidence ray tracing is the main modeling mode used in complex structure analysis, it is important to remember that the stacked section is assumed to be a high-signal, zero-offset section. Although this assumption is valid in the vast majority of cases, there are exceptions. Events from complexly deformed surfaces often do not have hyperbolic moveout and do not stack coherently, although they may be prominent in the synthetic section derived from normal-incidence ray tracing. Moreover, surfaces beneath a complex structure may be hidden within a shadow zone in a zero-offset model. (Shadow zones are discussed in Chapter 3 and an example is shown in Figure 40). In the real-data gather, and perhaps the stack, reflections from these surfaces may be visible because the complex structure may have been undershot at farther offsets.

Image ray tracing.—After stacking, data are migrated to place reflections in their correct lateral position. However, the algorithms most commonly used to migrate data, termed "time-migration" algorithms, leave a residual error caused by simplifying assumptions. Although these assumptions break down most readily when migrating data over a subsurface with strong lateral velocity variation they make the migration procedure computationally efficient and inexpensive. "Depth migration" algorithms do not make these assumptions and do not result in these errors. However, they are considerably more expensive. Larner et al. (1981) contrast the effects of time and depth migration. (The terms "time migration" and "depth migration" arise from the fact that the direct products of these algorithms are time-scaled and depth-scaled migrated sections, respectively. However, it is quite possible to obtain depth-scaled, time-migrated sections and time-scaled, depth-migrated sections as secondary products simply by applying vertical time to depth or depth to time conversion, respectively. The choice of scaling on the section does not necessarily imply the type of migration algorithm used.)

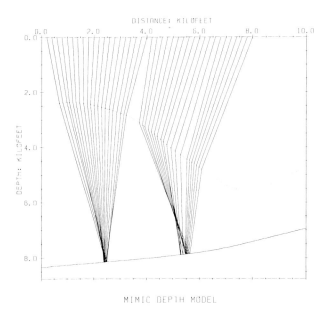

Fig. 20. Ray diagram showing two CMP gathers located at coordinates 2.1 and 6.1. Each gather contains ten rays with sources and receivers symmetrically positioned about the midpoint locations. Note that the reflection points within each gather are not identical. This phenomenon is known as CMP smear.

The error is predicted by a seismic ray termed the "image ray" (Hubral, 1977, see discussion in Sarvesam and Dhole, 1986). As will be shown, this ray is physically unrealizable. However, under certain circumstances the image ray accounts for the imaging results of time migration. The image ray is depicted in Figure 21. Consider a point source, or diffractor, located below a dipping reflector. The desired location of this diffractor after migration is vertically beneath A. Indeed, this would be the result if (a) a depth-migration algorithm was applied, (b) the overlying velocity field was accurately represented, and (c) if there were no sideswipe. Time migration, however, results in the error described by the path of the image ray. The image ray is defined as being normal to the ground surface. However, in contrast to the vertical ray, the image ray bends in accordance with Snells's Law as the ray propagates through the subsurface. Because of this raypath bending the location of the diffractor will be imaged beneath B and not A, and a time-migration error represented by distance A-B results. Note that the reflector along which the diffractor lies can have any orientation and, therefore, will not generally be perpendicular to the image ray. Because this zero-offset ray is, in general, not normally incident, the angle of incidence and the angle of reflection are not equal and the ray is physically unrealizable.

TIME MIGRATION ERROR

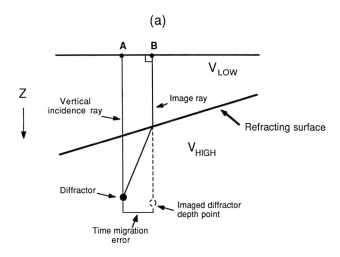

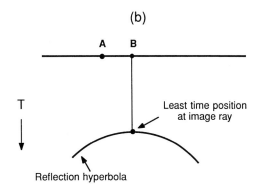

Fig. 21. (a) Schematic illustrating the time-migration error as represented by the image ray. (b) Diffraction hyperbola associated with arrival from diffractor. The image ray defines the least-time travel path from diffractor to recording surface.

This error results from the manner in which time migration algorithms work. Time migration algorithms consider the reflection as a continuum of diffractions and strive to collapse each diffraction hyperbola to its least traveltime peak. As shown in Figure 21, with laterally varying velocities and associated raypath bending, the least-time peak of the hyperbola is associated with the image ray and not the vertical ray. This is why image rays represent the outcome of time-migration.

An example of the application of image rays is not included among the case histories in this volume. However, Thorn and Jones (1986) presented an example from the North Sea.

Vertical rays.—Vertical rays simulate the depth-migrated section. These are pseudo-rays which (unlike real rays that obey Snell's Law) do not bend across interfaces. Therefore, depth-migrated sections represent the desired result in complex structure imaging. If there is no sideswipe, if the velocity field is accurately represented, and if events have been properly stacked, the depth migrated section will be the closest approach to a photographic image that seismic imaging can achieve.

When applied to a depth model, the vertical ray acts as a simple time-depth convertor and can be viewed as a continuum of one-dimensional seismic models. As a modeling tool these rays are most commonly used on time-scaled migrated data where the interpreter is evaluating velocity pull-up or sags.

INVERSE RAY TRACING

All of the ray tracing previously described begins with an estimate of the subsurface (model) and derives the seismic response. This is traditionally referred to as a forward problem in geophysics. It is also possible, using ray tracing, to do the opposite; to define the structure of an unmigrated reflection in time and directly derive the subsurface structure of the reflector in depth. This is traditionally referred to as an inverse problem (e.g., May and Covey, 1981). Figure 22 schematically shows the relationship between the two problems. Note that the inputs and outputs of the two are reversed.

The input in ray-trace inversion is a definition of reflection structure in either

THE FORWARD-INVERSE MODELING LOOP

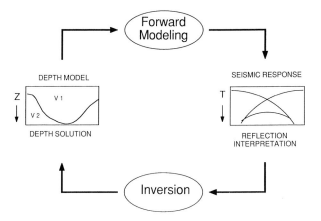

Fig. 22. The forward-inverse modeling loop. Forward and inverse processes are distinguished by their inputs and outputs.

two or three dimensions. "Reflection structure", as used in this text, refers to the shape of a reflection, in two or three dimensions as defined in time on either an unmigrated section, a time-migrated section, or a gather. "Reflector structure" refers to the shape of the reflector in depth as defined by a structure contour map.

Because reflection structure is the standard by which the plausibility of a model is judged, the reflection structure along with the layer velocity estimate, ultimately determines the final structural solution. The interpretive techniques, concepts, and problems associated with reflection interpretation are reviewed in Chapter 4. Because it involves the transformation of seismic observations (either zero-offset data or gathers) to a depth model, ray trace inversion is often referred to as migration. In fact, the outputs of ray-trace inversion and wave-equation migration of seismic traces ought to be similar.

As shown in Figure 4, 3-D ray trace inversion of a zero-offset reflection map is termed "map migration". The term "horizon migration" is used here for 2-D ray trace inversion. In addition, relatively new procedures have been developed to invert offset seismic data, for example shot or VSP gathers, to derive depth models. These techniques allow for derivation of subsurface structure if reflections are defined on offset gathers. They are termed tomography (see Figure 2.3) and are similar to procedures currently used in medical tomography.

Three main inversion strategies exist.

Direct ray-trace inversion

Direct ray-trace inversion is diagrammed in Figure 23. The input for the inversion is a definition of unmigrated reflection structure and a specification of layer velocities. Figure 23 shows that the inversion proceeds in a sequential rather than iterative fashion as in forward modeling. The structure of each reflector is successively derived and used to define the raypath bending that influences the derivation of lower reflectors. For this reason the procedure is sometimes referred to as a "layer building" method.

Zero-offset forward ray tracing, as described in the previous section, initiates at the reflector with the raypaths normal to the reflector. In direct, inverse ray tracing, the depth structure is treated as an unknown and, therefore, it is not possible to initiate rays normal to reflecting surfaces. Instead, the ray is initiated from the ground surface and continued downward. The departure angle of this ray is the angle at which the wavefront impinged upon the ground surface at the point where the ray was recorded. Figures 24 and 25 show how this angle can be derived from the local time dip of a reflection and subsurface velocity.

Figure 24 shows a curved reflection as the reflection would appear on a stacked, unmigrated section. The local time dip at point A is the dip (in terms of change in time over change in X) of the tangent to the curve at point A. Figure 25 shows two raypaths associated with a wavefront impinging on the ground surface. The difference in arrival times for the two raypaths over the separation distance Δx defines the local time dip. The product of this time dip and half the velocity equals the sine of the emergence angle of the ray. In inversion, when the ray is traced back into the subsurface, this emergence angle becomes the departure angle.

To reconstruct the raypath downward from the ground surface, the departure angle is derived as previously described and the raypath length is such as to account for the arrival time recorded at the point of departure. When such a raypath has been defined, the coordinates at the base of the ray are used to derive a reflecting surface that is normal to the ray at its terminus. Figure 26 shows how several of these rays can be used to define a sloping layer. In practice, rays are traced from points along the ground surface which are spaced apart by some user-defined interval. Also, if the program is interactive, the user is given an editing opportunity to delete raypaths that are spurious.

The above procedure defines the uppermost reflector in a model. To define a
second layer the same procedure is applied with one difference. As raypaths for
reflection 2 are traced downward they bend in accordance with Snell's Law as
they pierce the previously defined layer (or layers). In this way the derivation of
upper surfaces influences the derivation of lower ones. An unfortunate consequence of this is that positioning errors accumulate to the lowermost surface.
Figure 27 shows the rays bending as they pass through a previously defined
overlying surface.

Direct inverse ray tracing, like forward ray tracing, is readily generalized to
three dimensions. In the 3-D case, the ray departs normal to the local strike of the
reflection structure. In defining the surface, only the depth points at the base of

DIRECT RAY-TRACE INVERSION

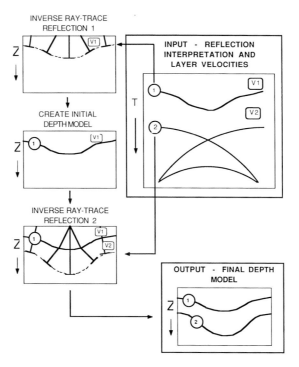

Fig. 23. A diagram of direct ray-trace inversion. The input (upper right) is in two parts. One
part consists of an interpretation of reflection structure. In this two surface example,
reflection 2 from the second surface is a multi-valued bow-tie reflection. The second part
of the input is layer velocities. The output is a depth solution which generally is a
geophysically compatible one; i.e., is capable of recreating unmigrated reflection structure.
The inversion operates in a sequential, layer by layer, manner as shown in this diagram and
in Figures 24–27.

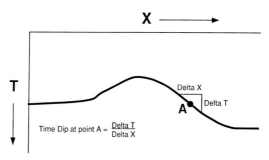

Fig. 24. Schematic diagramming the time dip of a reflection at a particular point.

the ray are displayed and editing is done in map view. More advanced programs attempt to utilize the local normals at the base of each ray in the final surface fitting.

Geometric inversion

A second inversion strategy does not involve ray tracing but simply relies on a geometric calculation that derives depth point locations from time dip, time value, and velocity. Each map is migrated independently. Each contour is migrated updip an amount that is appropriate for its time dip and average velocity.

This method is simple and, in fact, can be performed without the aid of a computer. The penalty paid for the simplicity is that raypath bending is not accounted for. With the advent of easy to use ray-tracing inversion programs, this procedure is becoming obsolete.

Ray-trace parametric inversion

The third inversion strategy is termed parametric inversion and is diagrammed in Figure 28. Note that the procedure is identical to that outlined in Figure 2

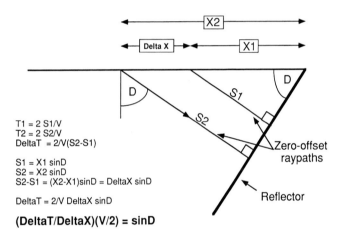

Fig. 25. Schematic showing how the departure angle (D) of a ray may be derived from the local time dip of a reflection and velocity.

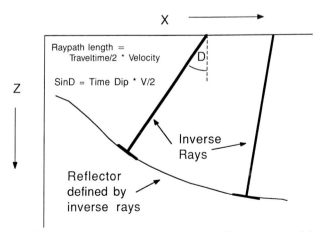

Fig. 26. Schematic showing how rays to the uppermost reflector in a model may be defined with inverse. Rays are downward continued from the ground surface at the calculated departure angle until the appropriate traveltime is accounted for. Reflector structure is defined such that the surface is normal to the ray at their terminus.

except that the comparison between the reflection interpretation and the synthetic data is an automated rather than an interpretive step. The exit from the routine can either be determined by (a) the number of iterations, (b) the reduction of error between the synthetic and interpreted reflections to a required tolerance, or (c) whether the error is continuing to diminish with further iterations. This procedure is regarded as inversion despite the fact that the procedure includes a forward modeling step. This step is included because the input is a reflection interpretation which outputs an earth model.

A key step in parametric inversion is the adjustment of the depth model which ensues from the synthetic/real data comparison. The various parametric inversion schemes differ in the numerical procedures employed in making these adjustments. All schemes seek to converge, in an efficient manner, upon a subsurface model whose seismic response closely replicates observed reflections. An initial estimate of the subsurface is needed for the first iteration. One measure of the quality of a parametric inversion scheme is how far off the initial model could be from the correct answer and still converge on the correct depth model. Ideally, the correctness of the initial model should not affect the ability of the inversion to converge on the proper solution. Only the reflection arrival times, and the layer velocities should determine the final result.

Figure 28 diagrams a procedure for the inversion of zero-offset data. Parametric inversion aimed at inverting offset arrivals, is termed tomography. These arrivals are generally interpreted on common offset sections, which have a similar appearance to stacked sections, rather than on the gathers themselves. Offset arrivals constitute a large body of observations; far too many to simulate by user-involved iterative forward modeling and not amenable to the direct inversion methods described previously. Hence, the need exists for a method which makes the synthetic/real data comparisons and model adjustments without involving the user. Examples of tomographic analysis may be found in Bregman et al. (1989), Bishop et al. (1985), and Chiu et al. (1986). Cutler (1987) presents an overview of the application and limitations of the technique.

Note that when inverting gathers, velocity information is an output rather than an input. In other words, the automated adjustments made to the model are made

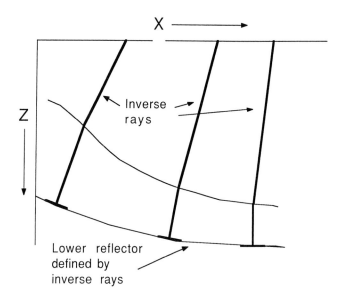

Fig. 27. Lower reflectors are defined in a similar manner as the uppermost reflector. However, as rays transmit through existing layers within the model they raypath bend across them. The ray is then continued downward with the new velocity.

**PARAMETRIC
INVERSION**

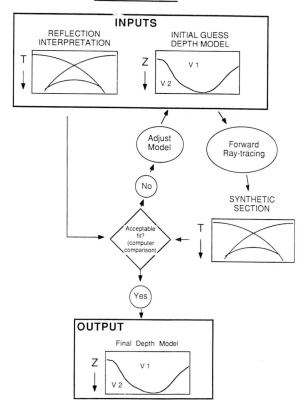

Fig. 28. Schematic diagramming parametric inversion. Note the similarly to the forward modeling diagrammed in Figure 28. The main difference is that the comparison between real data and synthetic output is made automatically without user involvement. Although this routine employs forward ray tracing the routine may be considered inversion because the reflection structure is considered as input and a depth model is output.

Table 2. Inversion inputs and outputs.

INVERSION I/O

	INPUTS	**OUTPUTS**
ZERO OFFSET RAY TRACE	1. Reflection interpretation of stacked data 2. Layer velocities 3. Initial Model *	Subsurface Model
OFFSET RAY TRACE	1. Reflection interpretation of gather data 2. Initial Model *	Subsurface Model

*   Parametric inversion

for both layer velocities and layer structure. In this way, the added constraints gained in modeling offset gathers, in contrast to stacked, data allows the velocity field to be derived rather than assumed. Table 2 outlines the inputs and outputs for zero-offset and offset inversion.

EVENT MODES

Ray-trace modeling is mainly aimed at simulating primary reflections. On occasion, structural interpretation issues arise which make considering other types of seismic events worthwhile.

Multiples

When an event is suspected to be something other than a primary reflection, it may be necessary to test a particular multiple scenario. Some modeling programs offer the ability to recreate a multiple with any combination of reflections from any surface. Figure 29 shows some multiple ray-tracing results.

Diffractions

Diffractions occur wherever reflectors are abruptly truncated. The most common association is with faults, but diffractions also occur beneath erosional surfaces and sometimes along the edges of fluid contacts. A diffraction event from the edge of a fault block is shown in Figure 30. The rays give the block edges the appearance of a point source. Indeed, as was shown in Figure 6, all reflections

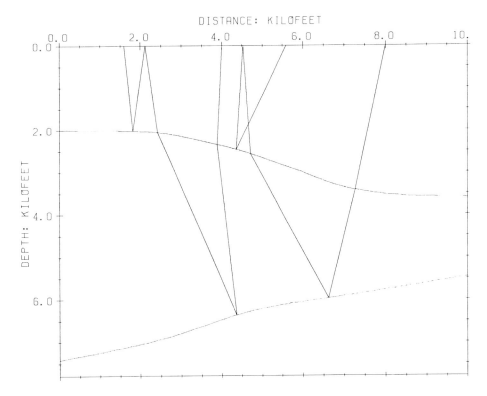

Fig. 29. Raypaths for a pegleg multiple reflection. (This is a multiple which contains an added bounce from a shallower horizon). Two raypaths are shown with sources at 4.0 and 8.0 kf. The rays have been traced with the intent of finding a multiple pathway that would be recorded at a location 2400 ft to the left of the source at 1.6 and 5.6 kf, respectively.

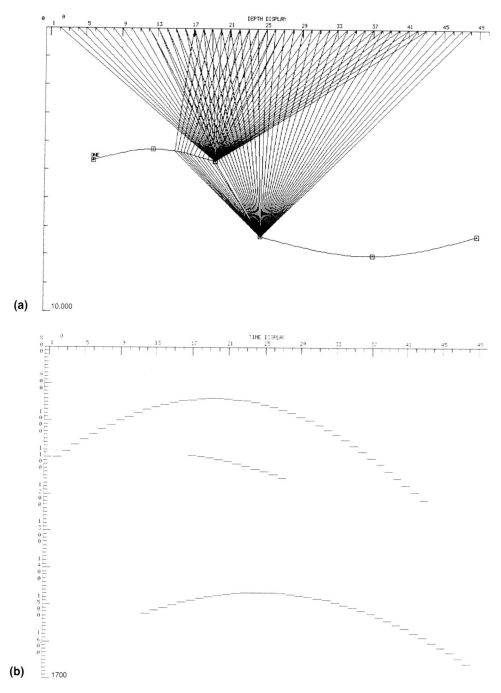

(a)

(b)

Fig. 30. Zero-offset raypaths (a) and arrival times (b) for a diffraction event. (a) The model consists of a single faulted surface, separating low- and high-velocity media, with the diffraction sources at the low- and high-side truncation points. (b) The arrival times show two diffraction hyperbola for arrivals from each point. There is also an event between shotpoints 17 and 29 which corresponds to a diffraction from the lower diffractor which travels partially through the underlying high-velocity media.

arise from a continuum of point (i.e., diffraction) sources which interfere to form a single continuous wavefront. What are often termed diffractions on stacked or zero-offset sections (and in this discussion) are the portions of these events which are incompletely interfered with because of reflector truncation. Because interference is incomplete, they remain visible on the stacked section.

In two dimensions, diffractors are depicted as point sources located at the point of truncation. Figure 30 shows a fault with two diffraction sources, one on the high side and one on the low side. Rays from these point sources take off at every azimuth and find a path to each receiver. The arrival time display in Figure 30 shows that the arrivals are hyperbolas with their peaks located at the trace position of the point source. (Recall from Figure 21 that, in fact, the peak or least-time position of the hyperbola is located at the image ray position. Had the model included some raypath bending surfaces above the diffraction point, this effect would be seen.)

In three dimensions, diffractors behave like line sources located along the trace of reflectors on the fault surface. For ray tracing purposes these line sources are difficult to define explicitly in three dimensions. However, because most software requires surfaces to extend across the model, the reflector is considered to take a sharp bend and continue along the fault. The corner formed by this bend has a small radius and acts as a line source.

Because diffractors are simulated by sharp bends in reflectors, the interpreter usually does not need to determine the nature of a particular event to be modeled. Whether considered as diffraction or reflection, the event should be found to emanate from a corner or shoulder of a reflector when modeled. However, accurate diffraction simulation is important when creating synthetic data sets to be processed with the intent of testing the imaging potential of a data process or geologic structure. In particular, the resolution of small fault blocks is to a great extent dependent on the ability to collapse a large portion of the diffraction event.

Refractions

Refraction studies have been used with great success in academic investigations of crustal scale structures. They are also used for defining the base of the weathering or low velocity layer. This information is used in static corrections to create a smoother time structure in the gather and on the stacked section. Ray trace modeling can be used to support both these applications.

Shear Waves

When a compressional seismic wave impinges on a reflecting surface at angles other than normal, part of the energy is "mode converted" to a S-wave. S-waves are important in studying stratigraphic effects because their amplitude response with offset is very different from that of P-waves, particularly when reflecting from porous media. S-waves may be important in investigating structural issues if they are mistaken for P-waves. As with the distinction between multiple and primary waves, modeling can be used to test whether a proposed event is a S-wave by attempting to recreate its arrival times.

CHOOSING THE PROPER RAY-TRACING MODE

When a decision is made to model a seismic section in order to solve a structural question; several other questions arise.

1. Should the problem be investigated by forward modeling or inversion?

2. Should the gather or the stacked section be modeled?
3. Should the unmigrated section be modeled (with normal- incidence rays), or the time-migrated section (with image rays)?
4. Should the modeling be 2-D or 3-D?

The first three of these questions are examined below. The last question will be examined in Chapter 3 on model building.

The type of ray tracing that should be done depends upon the issue under investigation. As stated in Chapter 1, seismic modeling of structures has three main applications. The first is to define structure from unmigrated reflections. This desire is presumably motivated by some perceived shortcoming in the wave equation migration that had been applied. Usually, this shortcoming relates to the inability of the migratin algorithm to properly handle raypath bending or, in the case of 2-D data, sideswipe. The second application is for modeling to act in a supportive role for data processing. The model fulfills this role by providing ray tracing displays and synthetic data sets. These outputs may be used to (1) establish VNMO trends, (2) locate key reflections on the unmigrated section, (3) design dip filter parameters, (4) specify the subsurface velocity field for migration, and (5) monitor imaging performance and resolution potential. The third application is to evaluate acquisition programs, mainly array length, spread length, and line length to ensure that reflections from all portions of the structure of interest are recorded.

Forward modeling versus inversion

Inversion procedures should always be employed when using ray tracing to define structure. Although forward modeling is an effective procedure for defining structure, inversion has the great advantage of making unnecessary numerous, iterative, modifications of the model.

In forward modeling the interpreter adjusts each surface, in succession from top down, until a satisfactory fit is achieved between simulated and observed reflections. In this way a geophysically compatible interpretation is derived. In contrast, inversion procedures derive structure directly (as described in the Inverse ray tracing section). This advantage is particularly strong when modeling in three dimensions. In three dimensions, adjusting a model, so as to bring the seismic response into conformity with real data, can be so onerous as to be impractical. In these cases, 3-D inversion, map migration, can be employed to derive a satisfactory model directly.

The output of inversion is a depth model. One could forward ray trace this depth model to see if the field data are recreated. At first glance, nothing would seem to be learned by forward ray tracing because the same rays that were traced in the inversion would be traced in the forward modeling. However, there are several good reasons for performing forward ray tracing after inversion.

First, and perhaps most importantly, forward modeling serves as a guide to interpreting reflection structure. The reflection interpretation can be progressively developed as the interpreter gains greater insight into the stacked section. In effect, the interpreter engages in an iterative procedure that consists of alternations of forward modeling and inversion as shown in Figure 31. An example of using forward modeling and inversion in this manner is given in Case History 9 by Fagin.

Second, inversion does not always yield a subsurface solution capable of recreating the unmigrated reflection. This result is particularly true when inversion is operating on multivalued reflections. (The interpretation, mapping and inversion of these reflections is discussed in Chapter 4). In these cases, depth points may overlap in either two or three dimensions if improper interval velocities are used. A surface fit to these overlapping points will probably be inaccurate. Forward modeling will call attention to this inaccuracy by showing that the modeled surface does not recreate the unmigrated reflection.

Third, forward modeling allows the interpreter to define those portions of the structure which are not constrained by the reflection map. The interpreter can alter those portions of the structure which are not sampled and still maintain the fit that has been achieved between synthetic and real data. Moreover, forward modeling can reveal which portions of a reflection have significant sideswipe, and thus where the interpreter may anticipate imaging problems.

Finally, forward modeling is more intuitively understandable than inversion, because inversion often has the appearance of a "black box". This difference can be important when one needs to communicate modeling results to those without extensive modeling experience.

In summary, inversion should always be considered when the object is to define a structure from reflections. Inversion works best when utilized in combination with forward modeling.

Gather versus zero-offset modeling

Modeling of gathers should be used to evaluate acquisition programs or processing procedures. It is possible to model gathers with the intent of defining structures, as was described in the discussion of tomography above. However, as previously stated, tomographic techniques are still experimental and time consuming. They should be used only where the interpreter needs to define structure with accuracy and where velocity information is limited.

Unmigrated versus time migrated section modeling

Because time-migrated sections can be simulated by image rays, they offer an alternative to modeling the stacked section. Figure 32 compares the two modeling procedures for 3-D inversion or map migration. Comparable diagrams can be envisioned for forward modeling and 2-D procedures.

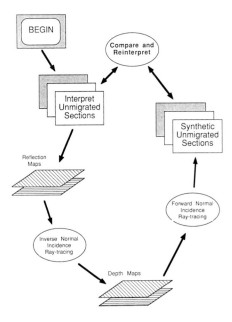

Fig. 31. A flow diagram showing how inversion and forward modeling may be used in combination.

There are two key advantages in using time-migrated maps. The first is that they are readily available because seismic interpretation is generally done on time-migrated sections. Viewed in this way the modeling procedure can be regarded simply as an adjustment made to a map that would have been generated in any case. In contrast, map migrating an unmigrated data requires interpreting unmigrated sections, which requires considerable additional work.

The second advantage is that the time-migrated section more closely resembles a photographic image than does the stacked section and, therefore, is easier to interpret. For instance, a syncline may appear as a bow-tie on the unmigrated section (and therefore might present some interpretational difficulty) but appear as a syncline, (perhaps with its trough displaced) on the time-migrated section. In general, reflections on stacked sections are more difficult to interpret than their time-migrated counterparts.

Against these advantages are three very significant disadvantages in image-ray modeling. The first has to do with 3-D effects. If 2-D migration has been performed on the data then image-ray modeling presumes that the seismic section contains the image ray as shown in Figure 33. To make an accurate time-migrated reflection map, only those portions of the seismic grid which are locally in good dip orientation can be used. This restriction could severely limit the available seismic control. In contrast, in making normal-incidence reflection maps from unmigrated sections, all portions of the seismic grid where the reflection is recognized may be used.

The second disadvantage is the assumption that the migration operation is properly simulated by image rays. However, the images resulting from different time-migration algorithms, employed in different manners are variable and cannot all be represented by the image ray (i.e., Yilmaz, 1987, for a description of the

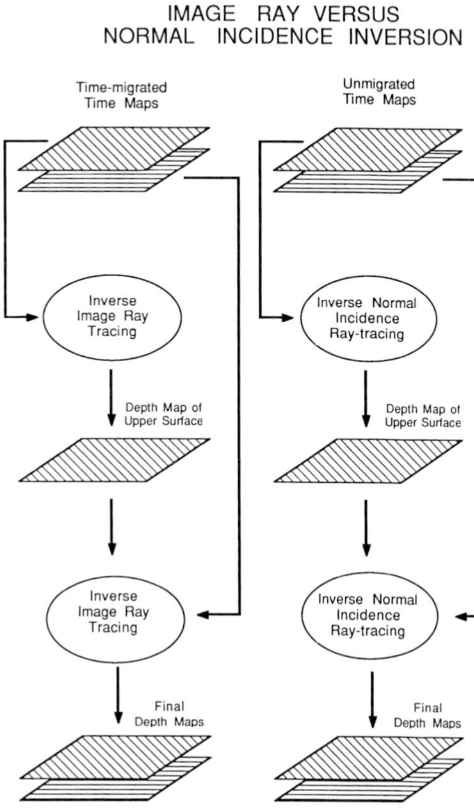

IMAGE RAY VERSUS
NORMAL INCIDENCE INVERSION

Fig. 32. Flow diagrams which contrast image ray and normal incidence ray inversion.

effects of different time-migration algorithms). In contrast, normal-incidence ray-tracing is performed on unmigrated sections and, therefore, issues concerning particular migration algorithms do not arise.

Finally, one must consider the effects of using one velocity field to perform the time-migration (generally some fraction of the stacking velocities) and a different velocity field to perform the image-ray tracing. These velocity fields are likely to be different because velocity fields for time-migration are commonly constructed from globally smoothed stacking velocities, while image-ray modeling utilizes horizon-keyed layer velocities. In contrast, velocities influence the unmigrated section in the stacking only. Different stacking velocities will impact the signal level of a reflection but probably not its reflection structure.

In summary, image ray tracing may involve some time savings and in certain cases may constitute the only interpretable data available. However, because normal-incidence inversion assumes much less about the input data the process is inherently more rigorous. If time is available and unmigrated reflections are interpretable, normal-incidence ray tracing is the preferred procedure.

RAY-TRACING DISPLAYS

Displays constitute the principal diagnostic output of a ray-tracing analysis, so the interpreter must understand what the displays are and how to use them. There are three types of displays to consider; displays of the model, displays of raypaths, and displays of the synthetic seismic record.

Displays of the model

There are three ways to display the seismic model; in map view, in perspective, or in cross-section (See Figure 34). The first two are available for 3-D modeling only. Map view and cross section are displays with orthogonal viewpoints, so they can be used to determine precise locations of structural features. Perspective diagrams are effective in rapidly conveying to the viewer the form of a structure,

Image ray lies within
plane of seismic section

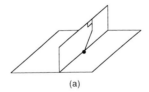

(a)

Image ray does not lie within
plane of seismic section

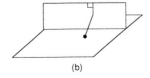

(b)

Fig. 33. Image rays both in (a) and out (b), of the plane of the seismic section. Time-migration operates in the manner represented by the image ray only in those cases where the image ray does lie in the plane of the seismic section. Whether doing 2-D or 3-D image ray inversion, the reflection interpretation should be made only from those portions of the seismic section in which this condition is honored.

so they are quite suitable for presentations. However, because they are not orthogonal projections, they are unsuitable for defining precise geographic locations.

Displays of the raypath

Raypath displays are used to determine the areas of the subsurface from which reflections are being recorded. The ability to create these displays is what makes

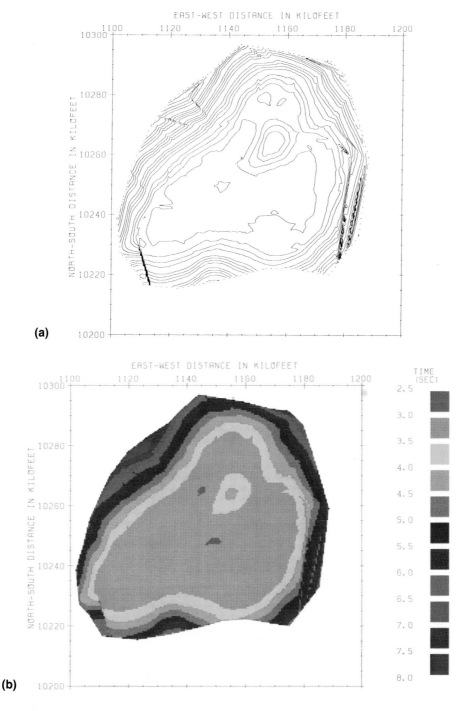

Fig. 34. Displays of the model; (a) line contour map, (b) color-filled contour map, (c) perspective diagram, and (d) cross section.

ray-trace modeling important to structural interpretation. The displays are critical in assessing sideswipe and raypath bending and they call attention to those portions of the structure from which reflections are not being recorded. Raypaths are displayed with the model in the background and so utilize the display types previously described. Figure 35 shows raypaths superimposed on a model. Note that although the perspective rapidly conveys a sense of where reflection points are located, the map view is more effective in showing their precise location.

The map view depicts several important raypath phenomenon. First, sideswipe is manifested on the display by those raypaths which diverge from the orientation of the seismic line. Second, raypath bending is indicated by those raypaths which are not straight. Bend points occur where the rays penetrate overlying surfaces. Third, some receivers record multiple reflections from this one surface, which are shown by several raypaths leading to one receiver location.

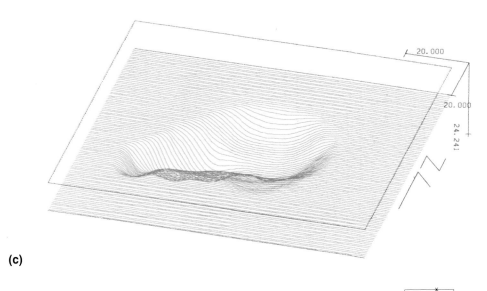

(c)

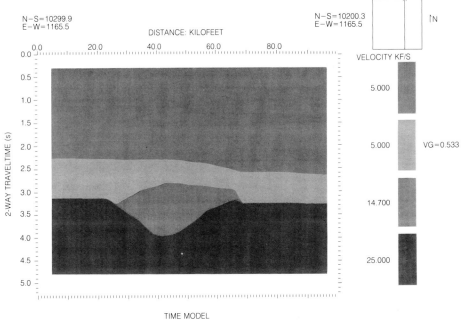

(d)

Fig. 34. Cont.

Displays of the seismic record

A record of reflection arrivals is stored at each receiver and these records can be displayed in the form of a seismic section. There are two types of displays. (See Figure 36.) The spike section indicates the times of each arrival with a spike. Spike lengths and polarity may vary according to the estimated amplitude of the reflection. Importantly, the spikes may be color coded so that the interpreter knows precisely which surface the reflection is associated with. This display should be compared to a reflection interpretation of the real data to evaluate the degree of conformity between the model response and the real data as shown in Figures 9–19. (Case History 9, pages 209–248). Because the accuracy of the model is measured by the models ability to simulate observed reflections, this is the most important display in the modeling process.

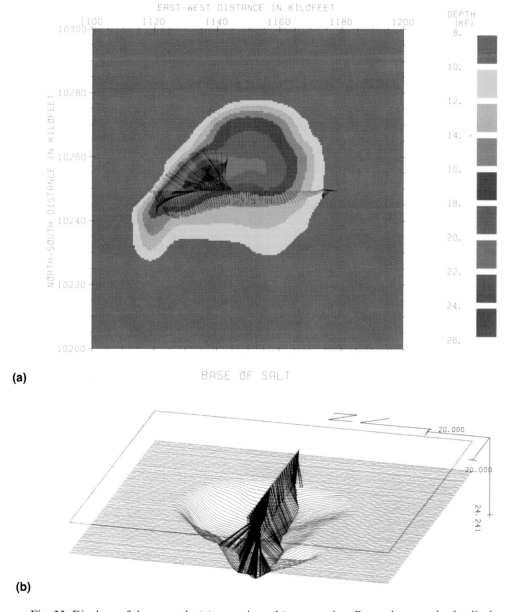

(a)

(b)

Fig. 35. Displays of the raypath; (a) map view, (b) perspective. Raypaths may also be displayed in cross section (see Figure 11).

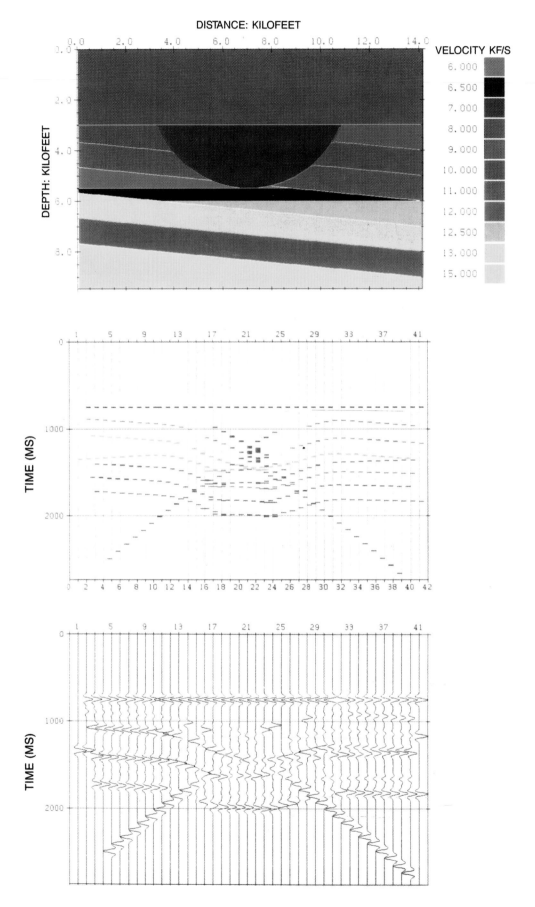

Fig. 36. Displays of synthetic seismic data. Upper figure is a ten-layered model including a low-velocity syncline. Middle figure is a spike display of arrivals color-coded for each surface. Lower figure is a wavelet display after convolving a wavelet with the spike series in each trace. Note that the wavelet displays make distinguishing separate events difficult.

Alternatively, the spike may be convolved with a user defined wavelet to yield a trace with a frequency bandwidth similar to the real data as shown in Figure 36b. Although this result may seem desirable, there is an important drawback. Because amplitudes from the different reflections are summed to yield a composite trace, reflections which are close together in time are no longer distinct. Problems interpreting the synthetic result may arise. Although both the spike and wavelet display may be generated, the interpreter soon discovers that all of the information bearing on structural issues is contained in the former. Using a wavelet with the proper frequency range is important if one intends to process synthetic data.

OTHER MODELING METHODS

Ray tracing versus wave equation modeling

As stated, ray-trace modeling works by representing the propagation of the seismic wavefront indirectly by the raypath. An alternative approach is to model the propagation of the wavefront directly, and this is the goal of wave equation modeling programs.

Wave equation modeling programs attempt to model the physical changes in the subsurface that are associated with the seismic disturbance. These changes, primarily pressure and particle displacement, are evaluated as the seismic wave is propagated through the subsurface. Commonly the subsurface is represented in a cellularized, rather than a layered, fashion with each cell assigned a variety of elastic parameters (some combination of P-wave velocity, S-wave velocity, density, Poisson's ratio, Young's modulus, or shear modulus).

For example, if the model is simulating a shot gather, then the seismic wave is initiated at the cell at the ground surface position of the shot. If the source is explosive, the wave may be characterized as having an initial compression followed by a rarefaction. The initial compression in this cell, after some time, excites a compression in adjacent cells and so the seismic disturbance spreads in the way envisioned by Huygen's principle. The velocity at which the wavefront spreads is determined by the physical parameters defined for each cell. Reflected waves initiate where the downgoing wave encounters an impedance contrast between cells. These reflected waves are directed back toward the ground surface where particular cells may be assigned geophone stations. These stations keep a record of the pressure or particle displacement changes occurring to the specific cell by sampling at a set time interval. This record constitutes the seismic trace, and the composite of all the geophones would constitute the simulated shot gather. This shot gather is the sole output of the modeling analysis.

The principle advantage of wave equation is in improved amplitude simulation and event continuity. This is particularly true in those cases where ray tracing routines exhibit difficulty in capturing and tracing rays along the edge of structures (Kennet and Harding, 1985).

Wave equation modeling can be used in the same way as ray trace modeling to solve structural problems; by deriving a model which creates a seismic section that simulates the real data. However, for structural problems wave equation programs have two significant disadvantages compared to ray tracing programs.

1. For a comparable subsurface they require far greater computer time. This added computer time arises from the need to keep track of the physical conditions in all cells at all times. This computer time requirement becomes so exorbitant that it generally prohibits 3-D modeling.

2. Wave equation programs do not result in displays which allow the interpreter to associate reflection points with shot and receiver points. The sole output of wave equation programs is the seismic record. There is generally no display which

depicts the propagation of the seismic wave through the subsurface. Therefore, there may be as many interpretation questions associated with the synthetic output as there are with the real seismic data. Such questions include the mode of a particular event, the specific reflector the event is associated, with or the portion of that reflector the event is associated with.

In contrast, ray-tracing programs create displays that answer these questions explicitly. Synthetic seismic sections can be made with spike arrivals color-coded or symbol-coded for different reflectors. Raypath displays show the interpreter which portions of a reflector associate with portions of the seismic section. Moreover, output raypath files are often created which contain coordinates for the beginning points, bend points, and end points of the raypath.

Despite these qualifications, there are circumstances where use of wave equation modeling programs to investigate structural problems is appropriate. In those cases where the interpreter seeks to assess the performance of an imaging process on a particular structure, amplitude simulation and continuity is important. An assessment of the best imaging strategy may require testing with synthetic data derived from a wave equation program. (An example is presented in Case History 4 by Morse et al.). If a modeler engages in this sort of investigation it would be done more effectively, if accompanied by ray trace modeling.

Physical modeling

Structures can be modeled using a physical representation of the model. The characteristics and advantages of physical modeling are described in Case History

Table 3. Stratigraphic versus structural modeling.

STRUCTURAL VERSUS STRATIGRAPHIC MODELING

	Structural Modeling	Stratigraphic Modeling
Seismic Phenomenon	Raypath bending	Interference effects
Data Parameters	Arrival times	Amplitudes
Processing Issues	Migration and stacking velocities	Pulse estimation, deconvolution and gain control
Model Parameters	Structure and interval velocity	Layer thickness and impedance
Size of Model	Extends vertically from ground surface to target. Wide enough to allow ray capture from target.	A narrow vertical window around target. No wider than target.
Units	Thick	Thin enough to account for interference effects
3-D effects	Important	Unimportant
Model Type	Depth	Time

7 by L. Zimmerman. Because the effort in building a physical model and collecting data over the model are so much greater than with a computer, the interpreter must have a clear understanding of the additional information that can be expected.

Structural versus stratigraphic modeling

Much seismic modeling is aimed at simulating seismic amplitudes for predicting porosity, lithology, or fluid type. This approach is termed "stratigraphic modeling" here, although the term "seismic stratigraphy" has been used elsewhere. Several examples are presented in Neidell and Poggiagliomi, 1977. The goals and methods of this type of modeling are different from structural modeling. These differences are outlined in Table 3. Both methods have in common the basic modeling goal: the simulation of the seismic section. However, because stratigraphic modeling seeks to simulate amplitudes, while structural modeling seeks to simulate arrival times, everything else about the two methods is different.

In stratigraphic modeling, interference effects are simulated which give rise to the observed amplitudes. Migrated sections are modeled using the 1-D convolutional model. In the convolutional model, amplitudes in the area of interest are affected only by the reflecting surfaces within a distance represented by the pulse. Geology outside of this zone need not be represented. In addition, the modeling can be done entirely in time, with the model constructed directly from the migrated section. The modeler need never go to depth, and indeed, there is often little incentive for doing so. Unless the modeler is considering the amplitudes of a sideswiped reflection, an unusual circumstance, there is no point in constructing a 3-D model.

The processing issues important to each modeling approach are very different. Velocity analysis, migration, and statics affect arrival times. Amplitudes are most affected by deconvolution, band pass filtering, pulse stabilization and estimation, and gain control.

Most exploration problems which require seismic modeling fall into one of these two categories but generally not both. (A possible exception is the simultaneous modeling of porosity within a reef and arrivals from its flank). When approaching a seismic modeling problem, the interpreter needs to be aware as to which of these two avenues should be taken.

CHAPTER 3

Model Building

The type of model that is built can greatly affect the quality of the simulation achieved. An improperly constructed model may fail to answer the questions at issue, incompletely simulate the raypath, or result in unnecessary effort. To avoid these errors, the following questions need to be considered.

Should the model be two-dimensional (2-D) or three-dimensional (3-D)?

How large should the model be?

Which and how many surfaces should the model contain?

Where should interval velocity information be obtained; and should interval velocities vary laterally or vertically between surfaces?

What level of structural detail should be portrayed in the model?

The answers to these questions depend on how they affect the main goal of ray-trace modeling: the simulation of the seismic raypath. Each of these questions are examined in turn.

2-D VERSUS 3-D DIMENSIONAL MODELING

In the last five years, with the advent of powerful computer workstations, the ability to perform interactive, 3-D, ray-trace modeling has become commonplace throughout the petroleum industry. This change in modeling capability represents a profound expansion of the modeler's ability to comprehend the seismic response to complex structures. Interpreters must now consider the question of whether a particular structural exploration problem is more efficiently evaluated in two or three dimensions. The disadvantage of 3-D modeling lies primarily in the added effort in model building, editing, and to a lesser extent, in ray tracing. This additional effort can be considerable; a modeling project which could be completed in several days in two dimensions, may take several weeks in three dimensions. Commonly, there is a limited amount of time that can be devoted to a particular project, and this consideration alone may govern the modeler's decision. An additional disadvantage of 3-D modeling is the memory requirement some software imposes. These requirements may result in limitations on the number of surfaces which can be modeled.

The primary advantage of 3-D modeling lies in its capability to simulate sideswipe. As discussed in Chapter 4, sideswipe is present in every complex structure grid and can cause profound but subtle imaging distortions, so this advantage is formidable. A subordinate benefit of 3-D modeling is the ability to model a series of lines with a single model. Additionally, a 3-D model allows the interpreter to view and evaluate a structural model by displaying a cross-section along any line of section and through any well control.

Given all these considerations, the decision as to whether to use 3-D modeling generally hinges on the sideswipe issue: Do dip lines over the structure show enough sideswipe to warrant the additional effort of 3-D modeling? Case Histories 4, 7, 8, 9, and 10 illustrate applications of 3-D modeling.

MODEL SIZE

Often, the deepest unit in the model should be the target horizon. Ray tracing to deeper units will not affect arrival times from, and therefore the structural definition of, the target horizon. However, there are times when the modeler may wish to model to a greater depth. Important examples are where velocity pull-ups and sags, occurring in reflections below the reservoir, are diagnostic of reservoir structure, porosity, or fluid type. In Case History 3, Lingrey uses a velocity pull-up, occurring on a reflection from a flat basement horizon, to constrain the number and thickness of overlying, reservoir imbricates. Alternatively, where reservoirs are porous or gas filled, a velocity sag appears on lower reflections. The position and amplitude of the sag can constrain pay thickness or porosity.

In addition, if the reflection from the target horizon is weak, then modeling a deeper, more prominent, reflection may be useful. This reflection choice will make comparing model results with the field seismic section easier. If the deeper reflector is thought to be conformable to the target horizon, then the structure of one could be inferred from the other. Also, the modeler may want to investigate deep reflections for structural control. For instance, in thrust belts, knowledge of footwall cutoff angles and positions may affect the way an interpreter defines the structure of a prospect in the hanging wall. The model should be wide enough to capture rays from any feature under investigation. Because the modeler must anticipate this need before building the model, schematic structures may have to be modeled. Additionally, to have the model extend outside of seismic control may be useful, if the modeler wishes to learn whether reflections have not been recorded, because of line length limitation.

SELECTING MODELING SURFACES

The decision as to which surfaces to include in the model should be guided by the need to satisfy both the interpreter's exploration objective and the modeling objective.

The exploration objective is the structural definition of those surfaces which affect the trap. Foremost of these is the "target horizon" which most defines the trap. Most commonly this target horizon is the top of a reservoir or the base of a sealing unit. Other surfaces which should be modeled are those which modify the trap, such as faults, unconformities, or the flanks of salt structures. The remaining surfaces which should be included in the model are those needed to satisfy the modeling objective: the simulation of the seismic raypath. If the target horizon is overlain solely by a layer of uniform velocity, and there is no raypath bending, then it would be unnecessary to include overlying surfaces in the model. This would be the case, for instance, in modeling the reflection from the water bottom.

In complex structures, the subsurface above the target horizon almost always contains variable velocity media that results in marked raypath bending. To properly simulate these raypaths, the model must include enough surfaces to portray the velocity variation that would account for this raypath bending. The number of surfaces depends on interval velocity variation among stratigraphic units, the complexity of structure overlying the target horizon, and the capabilities of the seismic modeling program.

In many stratigraphic sequences surfaces separate units of greatly contrasting velocities. Common examples are regional unconformities, the top and base of salt, and surfaces separating carbonate and clastic units. The abrupt changes in velocity across these surfaces are likely to produce strong raypath bending and they should be included in the model (Figure 37).

Some stratigraphic sections, typically fine-grained clastics, exhibit a simple, progressive increase in velocity with depth due to compaction. A familiar example is the Pleistocene and Tertiary section of the Gulf Coast. Figure 38 shows that in this sort of media the raypath will be curved and will steepen upsection. Although there is not a discrete surface representing a strong velocity change, a way must be found to model the curved raypath. There are two solutions. Some ray-tracing programs permit modeling layers with vertically varying velocities. If rigorously calculated, as in Figure 38, this modeling will result in a curved trajectory for the raypath passing through the layer. If the program does not allow for vertically varying velocities, then surfaces should be included that will divide the section into halves, thirds, or perhaps more, depending on the section thickness. In this way a bent raypath will simulate the curved path (see Figure 10). If these surfaces were sited along prominent reflectors, comparison of synthetic and real results can more easily be made. In Case History 10, by Reilly, these two approaches are compared.

In addition to these stratigraphic surfaces, there are structural surfaces which cause raypath bending. Faults commonly juxtapose units of different velocities. Modeling low-angle faults such as thrust faults and extensional glide plane faults

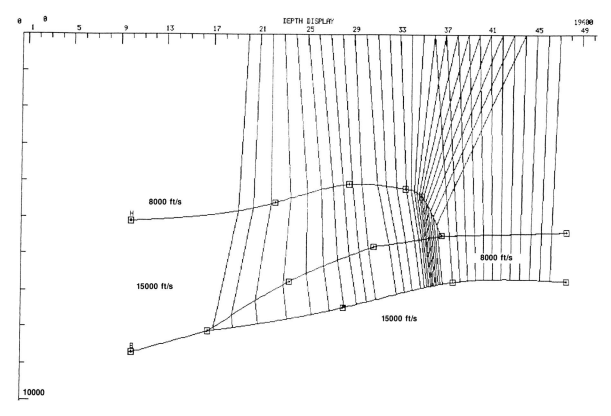

Fig. 37. Examples of raypath bending across the forelimb which forms the leading edge of a thrust sheet.

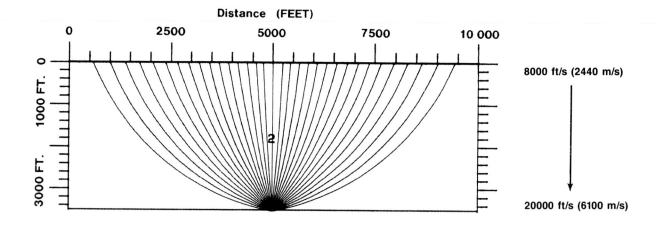

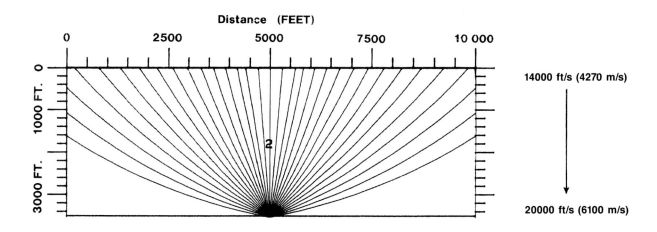

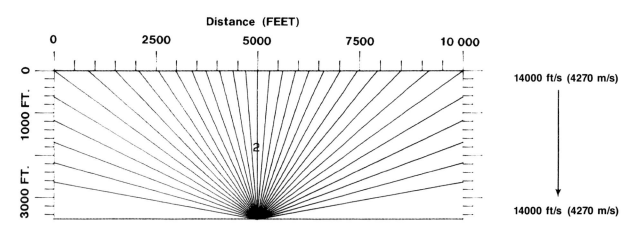

Fig. 38. Three models of a CDP gather with different vertical velocity gradients resulting in curved raypaths. Note that curvature increases with increasing gradient. If software does not permit the computation of curved gradients an alternative is to subdivide a layer, as was done in Figure 10. Figure courtesy of GMA Corporation.

is particularly important. These are often laterally extensive and produce raypath-bending effects over a large area. Both fault types can separate units with very different velocities. Thrust faults generally juxtapose older high-velocity units over younger low-velocity units, while extensional glide plane faults, such as those along the Gulf Coast, often juxtapose units whose velocity contrast is a consequence of difference in formation pressure.

Two strategies exist for representing faults in a model. They can be entered as discrete surfaces which truncate stratigraphic surfaces, or they can be represented as offsets in the modeled stratigraphic surfaces (see Figure 39). Entering faults as discrete surfaces is more rigorous but has the disadvantage of requiring more model-building effort. The advantages are first, the interpreter is allowed to model fault plane reflections, and second, velocity changes across the fault within the same stratigraphic unit can be more easily modeled.

Fluid contacts are obviously important in stratigraphic seismic modeling. They may also play a part in structural issues if the modeler wishes to understand, for example, velocity sag effects from an overlying gas sand. In these cases they should be included in the model.

Surfaces across which velocity inversions occur are particularly important to model. Raypath effects, such as defocusing and the development of shadow zones, are often difficult to foresee without the benefit of ray-trace modeling (Figure 40). The most common examples of velocity-inversion surfaces are the base of salt, base of carbonate, and thrust faults.

There is a limit to the thinnest unit that should be incorporated into a model. A thin salt bed might have strong raypath-bending occurring across its top and base, but both these bends will largely cancel each other and result in a negligible kink in the raypath without affecting the raypath direction (Figure 41). The thickness limit the modeler should consider is dependent on the velocity contrast of the thin bed and the modeling accuracy being sought.

Finally, most ray tracing programs restrict the complexity of surfaces in that they (1) require that modeled surfaces extend across the entire model, (2) prohibit surfaces that are multivalued in depth (in other words, surfaces which could be encountered twice by a single vertical borehole), and (3) prohibit crossing surfaces. Despite these restrictions, most structures can be represented in a model. Faults with reverse separation are represented by creating separate modeling surfaces for the high-side and low-side stratigraphy (See Figure 42). Each stratigraphic surface is linked to the fault as a modeling surface and continues to the end of the model. Surfaces with overhangs, such as diapirs and overturned folds, present a more difficult problem. These structures require an invisible surface, with no impedance contrast, extending from the point of vertical tangency to the end of the model (see Figure 43). Unfortunately, when represented in this way, the geologic surface has a discontinuous slope (i.e., kink) and contains a false diffraction source. The problem is magnified when attempting this representation in three dimensions. However, recent research advances, employing more innovative gridding procedures, may be the harbinger of a solution to this problem (See Figure 44, and Pereyra, 1988).

VELOCITIES

Along with surface structure, ray tracing models require the interval velocities between modeled surfaces. Interval velocities must be chosen to properly represent the traveltime through the modeled body.

For example, pure salt (halite) has a velocity of about 14 700 f/s. However, salt bodies may include other lithologies that have very different velocities. For

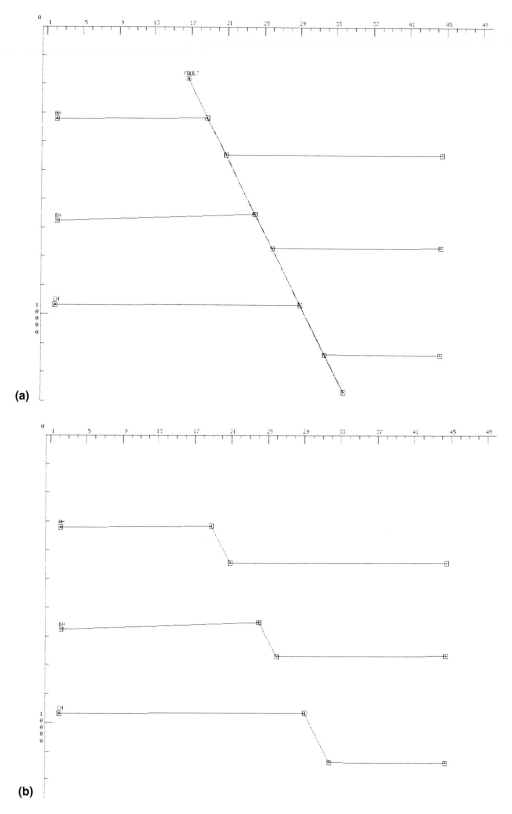

Fig. 39. Two ways of representing faults in a model. They can be depicted as (a) a discrete surface against which stratigraphic surfaces truncate, or as (b) an offset which links the same stratigraphic horizon in the footwall and hangingwall.

instance, caprock commonly encountered along the margins of piercing salt bodies may contain anhydrite or calcite. Also, bedded salt may contain interbeds of carbonates and siliciclastics. These included bodies of anhydrite, carbonates, or siliciclastics which can greatly alter traveltimes through the entire salt body. The modeler must assign interval velocities which properly represent traveltimes through the modeled layers.

Two issues to consider are the source of velocity information and how velocity variation is portrayed.

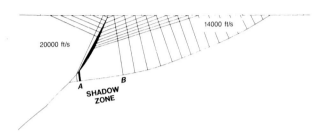

Fig. 40. A shadow zone is shown beneath a velocity-inversion surface which separates a high-velocity imbricate from an underlying low velocity imbricate. A "shadow zone" is an area of a reflector along which raypaths do not return to the surface. In this diagram, zero-offset raypaths are shown to the base of the 14 000 f/s layer. Raypaths are missing between points A and B because critical angle effects prevent the transmission of these rays through the overlying, high-velocity imbricate.

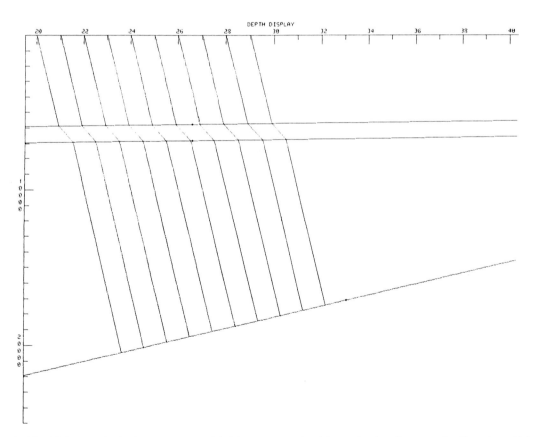

Fig. 41. A thin, high-velocity bed causes only a kink in the raypath without altering the raypaths direction. Although the top and base of the unit cause severe raypath bending their effects cancel one another. This bed should not be modeled.

Sources of velocity information

The two main sources of interval velocity information are well data and seismic gather data.

Velocities from well information.—Before considering well velocity information it is necessary to define two velocity terms: Average Velocity (VAVG) and Internal Velocity (VINT). The formula for average velocity is

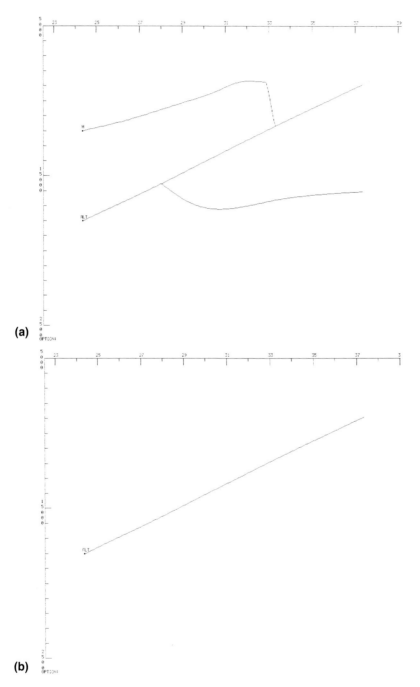

(a)

(b)

Fig. 42. A model showing how reverse offsets can be indirectly modeled with nonmulti-valued surfaces. In (a) a fault is shown which reverse offsets a stratigraphic surface. Although this stratigraphic surface is multivalued in depth it can be represented by three modeling surfaces as shown in (b), (c), and (d).

$$VAVG = \frac{(\text{depth to a reflector})}{(\text{vertical two-way traveltime of reflection})} \times (\text{two}).$$

VAVG is the velocity that relates the measured vertical traveltime from the seismic datum to a reflecting surface to the vertical depth to the reflector. A map of this velocity is commonly used to convert a time map made from a seismic grid interpretation to a depth map.

The formula for the interval velocity of a layer is

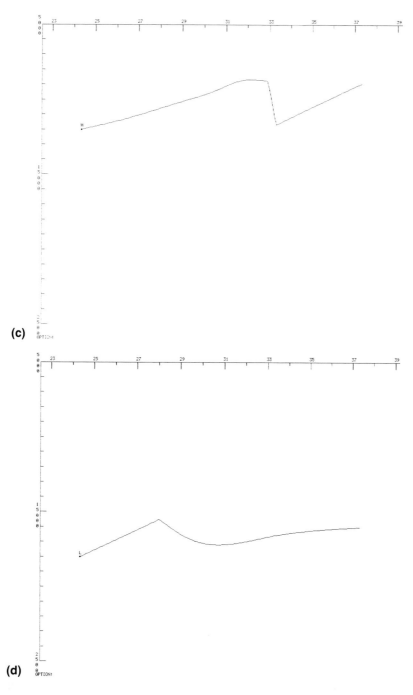

Fig. 42. Cont.

$$VINT - \frac{\text{(thickness of a layer)}}{\text{(vertical two-way traveltime through layer)}} \times \text{(two)}.$$

VINT is the velocity that represents traveltime through an individual layer Interval velocity maps are also commonly used for converting time maps to depth maps by summing the thicknesses of each layer to derive a depth to the bottom of the last layer. The thickness of each layer is derived by multiplying the observed traveltime through the layer (as measured on the seismic section) by VINT.

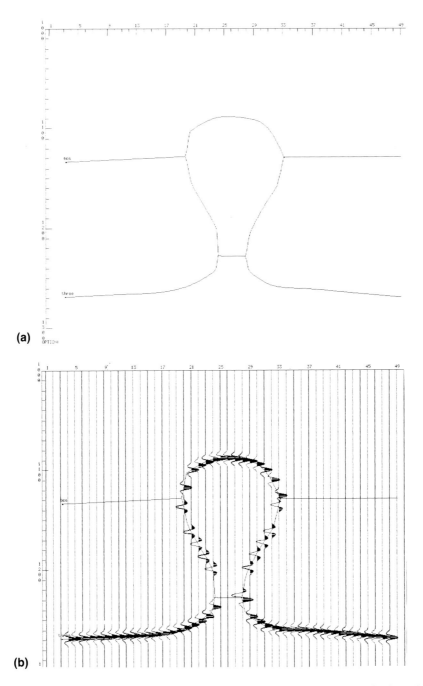

Fig. 43. A salt diapir can be represented using zero-impedance surfaces indicated in (a). They can be seen in (b), a vertical-incidence seismic section, but do not give rise to reflections. These zero-impedance surfaces serve to link different portions of the salt diapir in a way that allows the diapir to be portrayed without multivalued surfaces.

Because VINT predicts traveltimes through an individual layer, VINT is the velocity parameter that should be included in the seismic model.

With these definitions in mind, let's look at the three sources of well velocity information; sonic logs, velocity surveys, and vertical seismic profiles.

(1) **Sonic logs** are wireline logs which measure traveltime over some short interval, usually two feet. The sonic tool which is lowered in the borehole contains a multiple sources and multiple receivers at each end. The multiple sources and receivers allow the sonic tool to compensate traveltimes for borehole effects, primarily variations in the invaded zone and changing borehole diameter. Traveltime information is presented as an excursion curve measured in microseconds per foot, commonly on a scale of 40 to 140, and as integrated traveltimes shown as millisecond tickmarks. VINTs may be computed by adding the traveltimes between modeled surfaces and dividing the sum into the thickness between surfaces.

(2) **Velocity surveys**, or checkshot surveys, are created by shooting from sources located on the ground-surface into geophones lowered into the borehole. Traveltimes from source to geophones are determined by picking the first breaks at each geophone trace. Traveltimes from source to modeling surfaces can be determined by interpolating between geophones. However, significant errors can result if geophones are located far from these surfaces. Model layer traveltimes are derived by taking the difference in traveltimes from source to model surfaces. As with sonic logs, VINTs are determined by dividing thicknesses by traveltimes. Often velocity survey information is combined with the sonic log to form integrated sonic or velocity logs. In appearance these logs are similar to sonic logs, but both the log curve and integrated time tick marks have been adjusted so that traveltimes conform to velocity survey results. Curves which display discrepancies between traveltimes from the two data sets, termed drift curves, alert the interpreter to data problems. VINT's are derived from these logs as they would be from sonic logs. Because these logs combine the resolution of the sonic log with the gross (or low frequency) traveltime accuracy of the velocity survey, they generally provide better information than either data set alone.

(3) **Vertical Seismic Profile (VSP)** information is acquired in a similar manner to velocity surveys. The source is positioned on the ground surface and the receivers are in the borehole. The objective in VSP is not to measure traveltimes to geophones, although this information is contained in the VSP. The usual objective is to analyze reflections from surfaces below and perhaps lateral to the geophone to better understand the geology in the vicinity of the borehole. The entire trace of the VSP is analyzed rather than just the first break as in the velocity survey. In addition, to achieve the necessary vertical resolution, geophones are uniformly spaced at intervals of perhaps a hundred feet or less. Because VSP surveys can record reflections from steep interfaces, they are excellent tools for structural interpretation and modeling. Case History 5 by Nosal and others, and Case History 10 by Reilly present examples of VSP modeling. We may also consider a VSP survey as a high resolution velocity survey. VINT information is derived in a manner similar to velocity surveys with the benefit that because of the higher resolution, a geophone is usually located near each modeling surface penetrated by the borehole.

Velocities from seismic gather information.—In the absence of well information, VINT must be estimated from observations made in the seismic gather. In considering this approach, three additional concepts must be introduced.

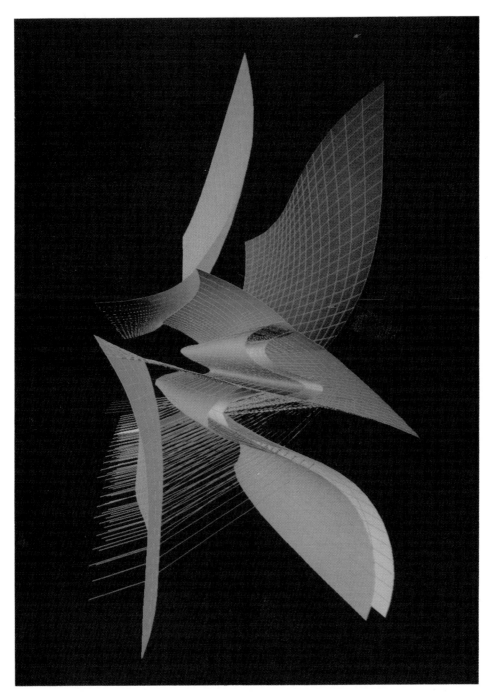

Fig. 44. Advanced gridding methods enable the representation of multivalued surfaces. (a) A thrust fault; courtesy of Weidlinger and associates, (b) a salt diapir; courtesy of the GOCAD project.

(1) **Moveout**—Moveout refers to the change in reflection arrival time with change in distance from source to receiver (offset). The formula for moveout for a horizontal reflector and a media of constant velocity is: (See Figure 45)

$$T_x{}^2 = T_0{}^2 + 4X^2/V^2 \qquad (1)$$

where

T_x = Reflection traveltime to offset x
T_0 = Reflection traveltime at zero offset
V = Velocity
$2X$ = Distance between source and receiver (offset).

This formula predicts arrival times at any offset for reflections from a horizontal reflector, given a constant velocity medium and zero-offset arrival time.

(2) **Root Mean Square Velocity** (VRMS)—The formula given for moveout is appropriate only for a single layer of constant velocity. The problem arises as to what velocity value would predict offset arrival times in the more common situation of a subsurface composed of layers with varying velocities. It intuitively would seem to require some measure of medium tendency of subsurface velocity such as VAVG. However, VAVG yields velocity values that are too low. Dix (1955) has shown that VRMS, a different measure of multilayer velocity behavior, can usually predict normal moveout in horizontally layered media to within a few percent. (The prediction would be exact except for the raypath bending across velocity interfaces).

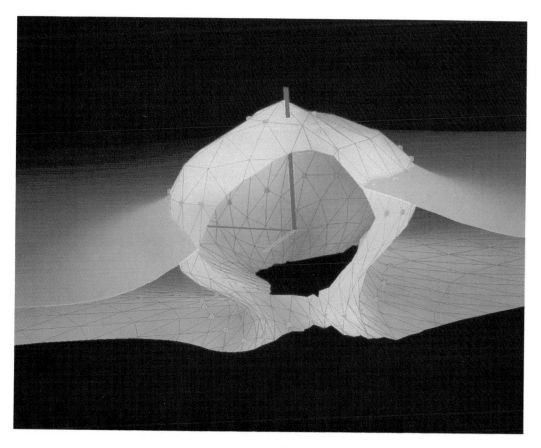

Fig. 44. Cont.

Equation 1 is modified so that the "moveout formula" for layered media becomes:

$$T_x{}^2 = T_0{}^2 + 4X^2/\text{VRMS}^2 \qquad (2)$$

where

T$_x$ = Reflection traveltime to offset x
T$_0$ = Reflection traveltime at zero offset
VRMS = Root mean square velocity
2X = Distance between source and receiver (offset)

The formula for root mean square velocity, often referred to as the "Dix formula", is:

$$\text{VRMS} = \left\{ \frac{\text{Sum } [\text{VINT}(i)^2 \times t(i)]}{\text{Sum } [t(i)]} \right\}^{1/2} \qquad (3)$$

where

VINT(i) = Interval Velocity of layer i
$t(i)$ = Time thickness of layer i

The formula can be restated to solve for VINT, given VRMS values for the top and base of the layer. Sometimes referred to as the "Inverse Dix Formula", the formula is:

$$\text{VINT}(i, j)^2 = \frac{\text{VRMS}(i)^2\, \text{T}(i) - \text{VRMS}(j)^2\, \text{T}(j)}{\text{T}(i) - \text{T}(j)} \qquad (4)$$

where

VINT(i, j) = Interval velocity of layer between surface i and j
$T(i)$ = Arrival time for a zero-offset reflection from surface i
VRMS(i) = VRMS down to surface i.

Equation (4) is a key formula for modeling applications because the interpreter is allowed to derive VINT from estimates of VRMS. Where dips are gentle, accurate estimates of VRMS can be made from an examination of moveout behavior in the seismic gather.

(3) **Normal moveout or stacking velocities** (VNMO)—Equation 2 implies that examining a common-midpoint gather to see which VRMS value best predicts observed moveout is possible. This procedure, termed a "velocity analysis", is done by summing amplitudes in a gather along the moveout path predicted by a series of VRMS values. If a reflection corresponding to a particular VRMS value exists in the gather, the summing will be in phase (amplitude values of the same sign will be summed) and the sum amount will be large. If there are no reflections

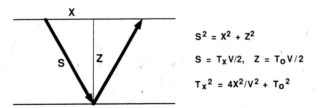

Fig. 45. Derivation of reflection arrival time (T$_x$) as a function of source-receiver offset (2X) for a flat bed.

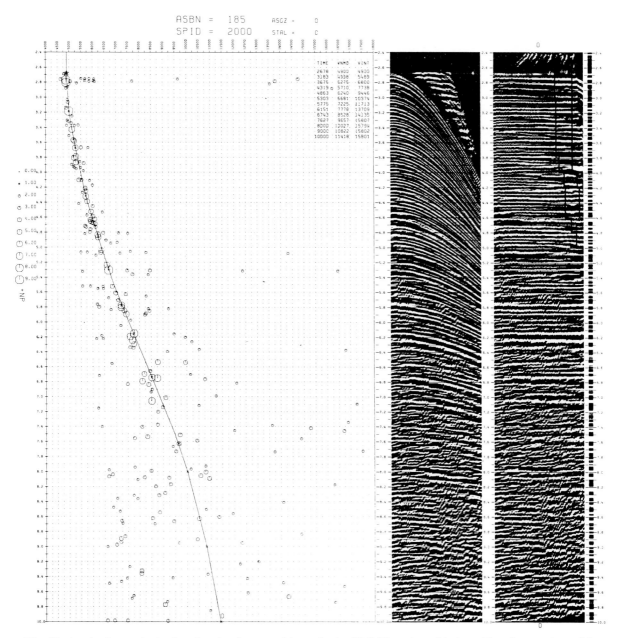

Fig. 46. A velocity analysis plot showing how stacking velocity VNMO varies with time. The plot consists of three parts, all three of which have time as their vertical axis extending from 2.4 to 10.0 s. The center panel is a seismic CMP gather with zero source-receiver offset on the left. (Notice that, as expected, the arrival times of reflections increase with increasing offset). At far left is a "semblance plot" with the horizontal axis being VNMO and extending from 4500 to 18 000 ft/s. The octagonal symbols on the plot represent those VNMO values (at particular times) for which there are peaks in the summed amplitudes. (The larger octagons represent the greater sums). Notice that VNMO continually increases with increasing time. Also note that where the data have low signal below 7 s, the velocity trend is difficult to define. The line drawn on the plot is the velocity function defined by a velocity interpreter. In the upper-right hand corner of the semblance display is a listing of the time/VNMO points picked by the velocity interpreter and the VINT values derived using the Inverse Dix formula (3.4). At far right is a "NMO-corrected" or "spread-corrected" gather showing the flattening of reflections achieved by applying the velocity function.

corresponding to this VRMS, then summing will be out of phase and the sum value will be small. Figure 46 shows the results of one analysis. The most common use of this information is to define the pathways along which gather data will be stacked to form the zero offset section. When VRMS is estimated in this way, the result is variously termed a "stacking velocity", a "moveout velocity", or VNMO.

Because VNMO values are the outcome of the velocity analysis which precedes stacking of multifold seismic data, they are almost always available to the interpreter for defining modeling interval velocities. If VNMO values are located along modeling surfaces, then an interval velocity can be derived directly by using the Inverse Dix Formula (Equation 4). If not, then VRMS values will have to be interpolated to appropriate positions, or velocity analysis plots may have to be reanalyzed.

Under what conditions are stacking velocities good estimators of VRMS? Stacking velocities are derived using the Dix formula which assumes a subsurface without dip. They would, therefore, seem unsuitable in complex structure environments and, indeed, VRMS estimates can deteriorate rapidly when structural dip varies significantly in a section. For example, Figure 47 shows reflection travel paths from a vertical reflector for a common midpoint gather. The figure shows that, for a vertical reflection, the traveltime of the reflection does not vary with offset. The only VRMS value which predicts invariant moveout has a value of infinity. Clearly, VNMO in this case is not an estimator of VRMS or any other sort of real subsurface velocity.

Zero-offset case

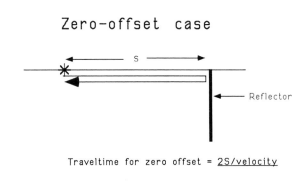

Traveltime for zero offset = 2S/velocity

Offset case

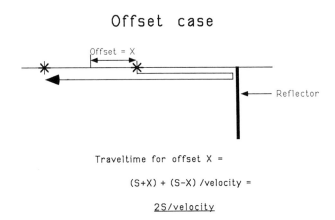

Traveltime for offset X =

(S+X) + (S-X) /velocity =

2S/velocity

Fig. 47. Reflections from a vertically dipping reflector indicating that arrival times do not vary with offset for this case. Such a moveout pattern would result in a stacking velocity (VNMO) of infinity.

If, in complex structure environments, VNMO values do not represent intrinsic subsurface velocities, how should the interpreter view them? They should be regarded simply as labels for the hyperbolas which best stack reflections in common midpoint gathers. As such, they cannot be used directly to define interval velocities for modeling. However, stacking velocities hold the potential for constraining interval velocities in some fashion.

To this end, there are several ways to utilize stacking velocities.

(a) Utilize VNMO values obtained in off-structure areas where there is gentle dip.

(b) Utilize VNMO values obtained on strike lines. A CMP gather oriented perfectly along strike is identical to one taken over layers with zero dip with one difference; the plane containing the raypaths is tilted away from the vertical by the angle of dip (Figure 48).

(c) Attempt to account for the dip effect using the formula from Levin (1971):

$$VNMO = VRMS/(cosine\ dip\ angle). \tag{5}$$

This formula predicts moveout over a dipping horizon if there is no raypath bending. If overlying geology is moderately complex, the formula should not be used. In the vertical reflector example of Figure 47, VNMO goes to infinity because the cosine of dip is zero.

(d) Although VNMO values over complex structure are not measures of intrinsic subsurface velocities, they are a consequence of subsurface velocities as well as subsurface structure. Therefore, they can become the object of forward modeling. In Case History 6, by Zijlstra and others, a procedure is presented in which an initial model is gather ray traced and perturbed (both layer structure and layer VINT) until calculated T-zero and VNMO values simulate observed values. In this way VNMO values are made to constrain VINT without making the Dix formula assumptions. Another example of using VNMO values in this way was presented in Gerritsma (1977).

(e) Gather data can be transformed in such a manner that VNMO values derived on the output gather more closely resemble VRMS. In effect, the dip factor is removed. This process is termed dip moveout or partial migration. However, dip moveout is a process which does not properly account for raypath bending associated with lateral velocity changes. VNMO values should still be treated with caution if the overlying structure is complex.

(f) In conjunction with prestack migration, a focusing velocity analysis can be done. This analysis scans a range of velocities and determines where amplitudes peak. Because migration algorithms seek VRMS velocities to best collapse diffractions, these values should be unaffected by subsurface dips. If the prestack migration is also a depth migration procedure, then raypath bending associated with lateral velocity changes should be accounted for; if not, this is an important limitation on the results. If the analysis is 2-D, another limitation is that the line must be in dip orientation.

Of these methods, the first two are likely to be easily accessible to an interpreter analyzing a grid of seismic data. The last four require special analytical effort but have the potential for yielding more detailed and accurate information.

In summary, the two data sources available from which the modeler can obtain velocity information are well and seismic. Well data are more accurate, but seismic data are more ubiquitous. The next issue to consider is how and if the modeler should portray vertical and lateral velocity variation.

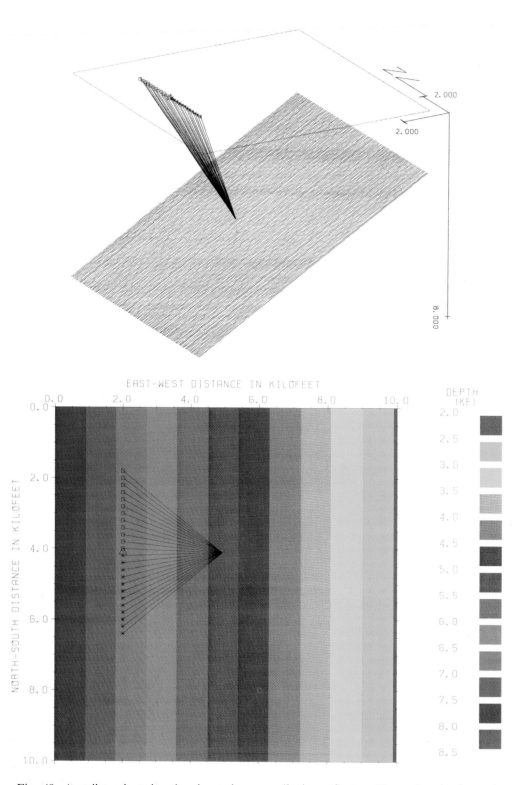

Fig. 48. A strike oriented gather located over a dipping reflector. The gather is shown in perspective (a) and map view (b). The raypaths in the gather are identical in appearance to those acquired over a flat horizon and subsequently rotated from the vertical plane. They, therefore, would be suitable for inferring interval velocities.

Representing velocity variation

With the exception of certain relatively homogeneous media such as seawater, salt, or granite, geologic layers will often contain enough velocity variation to warrant consideration in model building. Most ray trace software allows modeling of laterally varying layers and some allows vertical variation within a layer as well. If the modeling is being done over a producing field or a mature exploration area, there may be enough well control to map lateral velocity variation. However, often there is insufficient data to make meaningful maps.

One way to define variable velocity layers is to employ a geologic model. For example, if an areal facies change from carbonates to clastics is known to occur, this information can be employed in making an interval velocity map. Similarly, vertical velocity gradients in clastic sections are a function of maximum depth of burial and, if uplift has not occurred, can be modeled as a function of present day depth. The structure map of the base or top of a layer can be used as a velocity variation map by substituting contour velocity values for contour depth values.

LEVEL OF STRUCTURAL DETAIL TO MODEL

In building the model the interpreter must decide how much structural detail to incorporate in the model. In other words, how narrow a fault block and how small a fold should be represented in the model? This question summarizes to (1) the tradeoff between the added effort in representing small features versus the benefit in added structural resolution, and (2) the resolution limitations of the seismic data.

Both of these issues determine the grid size to use in a model. Model surfaces are represented in a digital computer by a grid, similar to the grid drawn on a sheet of graph paper. Each grid cell has the same size and shape (square or rectangular) but each contains a different elevation value. In this manner any continuous surface can be represented by a grid. Grids are constructed from randomly spaced control points which generally follow the trace of contour lines from a digitized contour map. Ray tracing computation time increases exponentially as the size of the grid cell decreases linearly. There is, therefore, a computer time cost in choosing too small a grid cell size. However, if the grid cell is too large, to model a narrow fault block or fold may not be possible. To properly represent slopes in a fault block requires at least three grid cells. To properly represent the shape of a fold requires from six to eight cells per wavelength. To choose the proper grid size the size of the smallest structure to model needs to be considered. Modeling small fault blocks and folds can involve considerable effort, particularly where they are represented by multivalued reflections. In deciding whether such effort is warranted, the interpreter must consider the resolution limitations of the seismic data; not only the theoretical limits on spatial resolution but the broader limitations of the interpreter's ability to correctly infer the reflection geometry of small structural features. Often the best procedure is to begin modeling (or inverting) relatively simple structures. Ray tracing results from these initial steps should guide the interpreter to successively more detailed solutions.

CHAPTER 4

Reflection Interpretation

The goal of modeling is to elicit the seismic response of a subsurface structure. The plausibility of that structure is gauged by how well the form of the simulated reflection matches observed reflections. Correctly interpreting and mapping reflection structure, as opposed to depth structure, is therefore critical to the modeler's ability to assess his results and derive the proper structural solution.

This chapter examines the meaning of reflection maps, shows how a reflection is organized in three dimensions, and explains how the map is constructed from various manifestations of the reflection, including diffractions and sideswipe. It also shows how the reflection map can be migrated in three dimensions, using inverse rays, to derive a depth structure which properly accounts for raypath bending and sideswipe. The procedure is termed map migration.

THE ZERO-OFFSET MODEL OF A REFLECTION MAP

Chapter 2 reviewed how reflections on stacked, unmigrated seismic sections are intended to simulate the way they would appear if they had been acquired using a single-fold, zero-offset system. In this conceptual acquisition system, each trace is created by a single receiver recording reflections from a single source located immediately next to the receiver.

Consider this acquisition system in three dimensions. The field system consists of a source bolted side-by-side to a receiver (see Figure 49). (Assume that we have devised a way of preventing the source from disrupting both the receiver and the data gatherer). We've also managed to make the survey lightweight and portable so we can take the system from station to station on the ground surface much as is done in a gravity survey.

Figure 50 shows the geology of the field area which consists of a single reflector dipping to the right. At each of the stations shown on the ground surface, we activate the source and propagate the seismic wave to the reflector and back to the receiver. Because this is a zero-offset system, the raypaths are normally incident at each reflector, and the downgoing and upgoing raypaths are identical.

At each station we note the arrival time of the reflection (Figure 51). We can make this measurement in several places until we have enough data points to define the structure to the detail we desire. By contouring the arrival times, as in Figure 52, we produce a "reflection map" which depicts the two-way arrival time from the reflector that can be expected along a seismic trace located at any point in the map area.

In reality, seismic data are collected along lines which correspond to ship tracks or perhaps vibrator truck paths. However, this acquisition simply represents a sampling difference and the meaning of the map remains the same. Any seismic

Fig. 49. A conceptual acquisition system with source (S) and receiver (R) located in the same position.

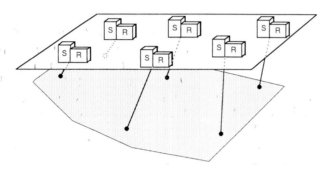

Fig. 50. Operating the acquisition system at several places.

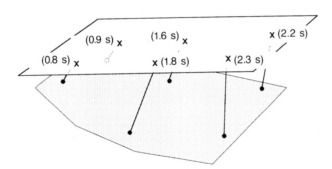

Fig. 51. Obtaining arrival times at each location.

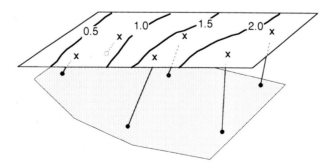

Fig. 52. Constructing a reflection map.

line through the area can be viewed as a linear collection of sample points across the reflection map. The reflection, which has yet to be migrated, is depicted on the seismic section as though the reflection were a cross section through the map (Figure 53).

Another way to view the reflection map is as a snapshot of the wavefront from an exploding reflector as the wavefront emerges at the ground surface (Refer to the discussion in Chapter 2). In this conception, the subsurface reflector is seen as an array of point sources, uniformly distributed over the reflector, which are set off at the same moment. At any point along the reflection surface, the local time dip (when combined with a subsurface velocity) indicates the angle at which the wavefront impinges on the ground surface. As was shown in Chapter 2, this concept is important when considering how inverse rays are traced.

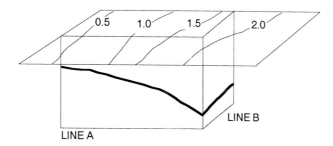

Fig. 53. Seismic Lines A and B can be viewed as a cross section through the reflection map.

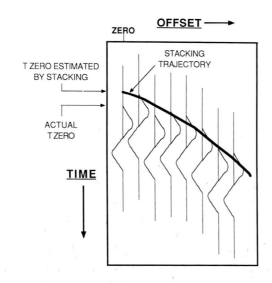

Fig. 54. The zero-offset arrival time is not always the same as the stacking position of a reflection.

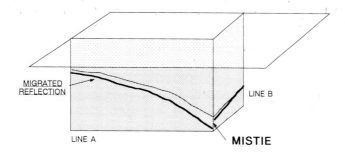

Fig. 55. Seismic lines A and B mistie after migration.

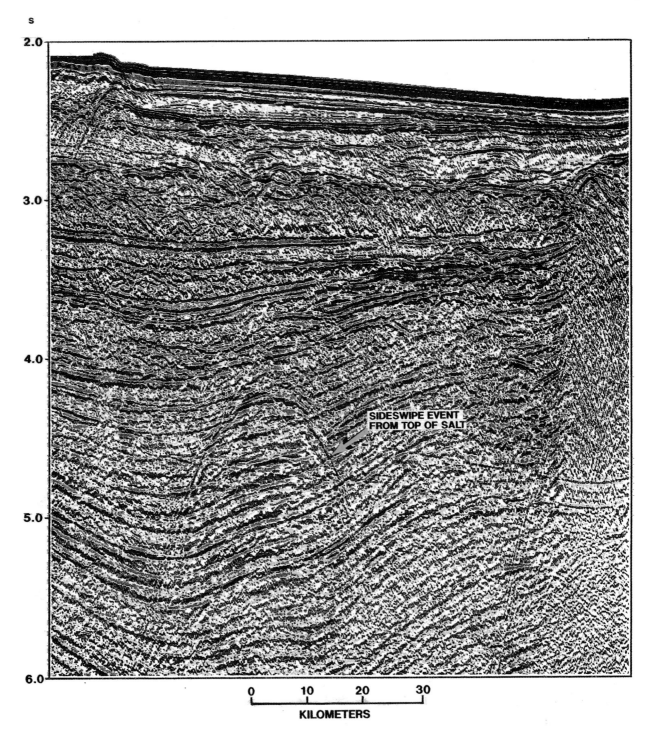

Fig. 56. (a) Seismic section displaying sideswipe event distinguished by superimposed image of a salt structure on throughgoing sedimentary section. (b) Perspective raypath diagram of this line showing the direction of sideswipe.

CONSTRUCTING THE REFLECTION MAP

Although the concept of a reflection map is straightforward, its construction requires some additional thought and effort on the part of the interpreter. In particular, the interpreter must wear a somewhat different hat in that the seismic section is no longer viewed as the "photographic image" with the object of depicting geologic structure in cross section. The section is now viewed in accordance with the concept shown in Figure 53; as a cross section through the reflection map depicting the recorded reflection.

In interpreting a seismic section, normally the interpreter can infer those portions of a structure not imaged on the section. However, in interpreting reflections for the purpose of constructing a reflection map, one should be less aggressive. Only valid observations should be mapped for modeling or inversion. Estimating *in depth* those portions of a structure that are unconstrained is easier, than estimating their seismic response. In concert with this consideration, inversion procedures for reflection maps, known as map migration, do not require reflection maps to extend across the entire modeling area.

With these ideas in mind, we now consider some of the important interpretational issues which exist in constructing the reflection map.

Tying unmigrated reflections

In Figure 53, consider the seismic trace recorded at the station located where Lines A and B intersect. Clearly, if this trace is recorded twice, as part of each line, the two traces would be identical. This result would be true because the identical experiment has been conducted for both lines. Therefore, we would expect Lines A and B to tie at their intersection. Unmigrated lines tie at their intersection because they imply identical acquisition conditions there. One benefit of reflection mapping is that these conditions are generally the case, whereas, when tying migrated data, as shown later, it is not always so.

On occasion, unmigrated reflections will not tie through a seismic grid. The causes for this lack of tie must be related to reasons why the seismic experiment at line intersections is in reality different for different lines. Common examples follow.

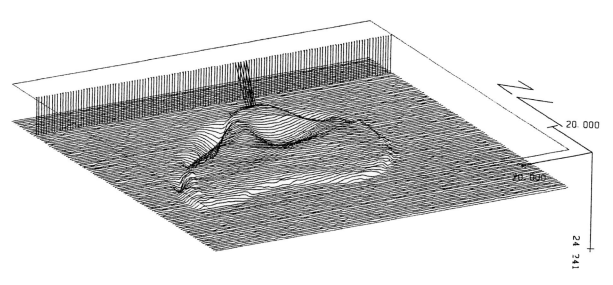

Fig. 56. Cont.

(1) Different statics assumptions, such as those concerning static datum, static velocity, and weathering layer thickness, will cause differences in reflection arrival time at line intersections.

(2) Differences in the basic wavelet from line to line may cause phase shifts when tying reflections. Wavelet differences may be related to source type, source strength, data filtering, or deconvolution.

(3) Stacking of reflections which have non-hyperbolic moveout causes differences between the zero-offset arrival time and the stacked T-zero position (Figure 54). Different lines might stack to a different *T*-zero position at the same trace location.

(4) As mentioned previously, steeply dipping events cannot be stacked simultaneously with flat reflections. If the choice of reflection to be stacked is changed from line to line, then reflections will not tie. We seek to relieve this problem with dip moveout processing.

(5) If there are differences in signal-to-noise content between lines, weaker reflections may appear on the higher signal line but not on the weaker one.

If unmigrated seismic lines are expected to tie, what should be expected of migrated seismic lines? Figure 55 illustrates what occurs to the reflections on lines A and B when the reflections are subjected to a 2-D migration operator. Line A is in good dip orientation with respect to reflection structure. As we would expect, when migrated, the right-dipping reflection steepens. Note that, with steepening, the reflection is delayed in time by an amount delta t. In contrast, the reflection, appearing on Line B, has little dip and is not greatly affected by the migration operator. Therefore, after migration the two reflections have different arrival time values at the line intersection and there is a mistie of about delta t. An example of such a mistie is presented in Case History 9, Figures 3a and 3b.

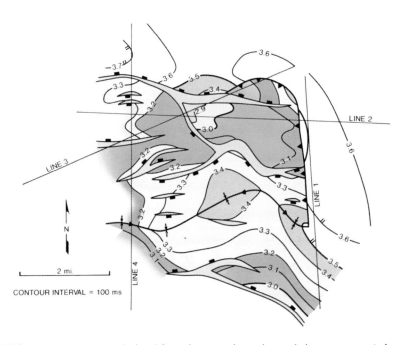

Fig. 57. Time structure map (derived from interpreting migrated time structure) drawn on a "Horizon A".

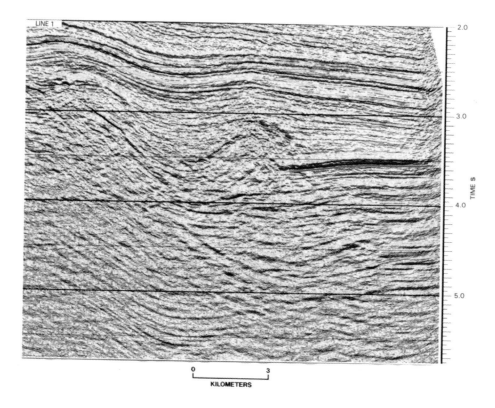

Fig. 58. Migrated Line 1 showing two events from Horizon A in reverse offset.

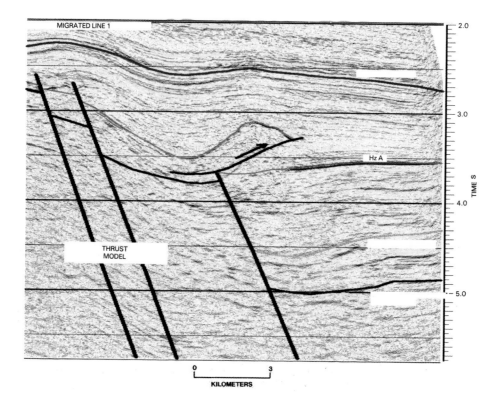

Fig. 59. Interpretation of Line 1 accounting for the reverse offset in Horizon A with a thrust fault.

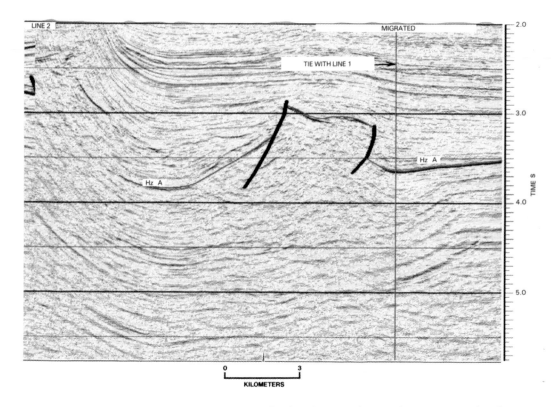

Fig. 60. Migrated Line 2 indicates that the high side reflection from Horizon A does not extend as far eastward as Line 1.

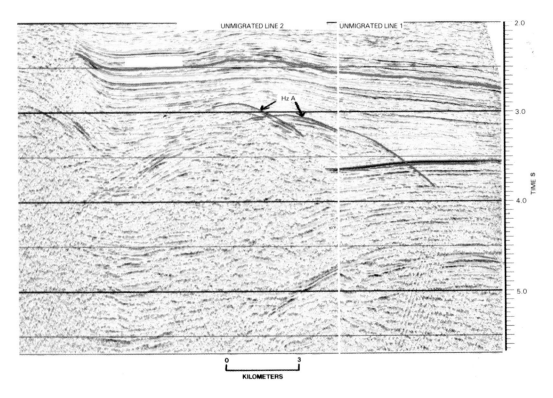

Fig. 61. Tie between reflection events on unmigrated Lines 1 and 2. The tie demonstrates that the two events are the same reflection.

Sideswipe

In constructing the reflection map, being cognizant of the three-dimensional aspects of reflection structure is important. Key to this awareness is an understanding of the phenomenon of sideswipe and how sideswipe affects reflection structure. Without this understanding, the interpreter will be unable to properly correlate a reflection through a 2-D grid, unable to prepare a correct reflection map, and ultimately, unable to properly define depth structure in three dimensions. Sideswipe is described and several examples are shown in this section.

A sideswipe reflection is one whose reflection points lie outside the plane of the seismic section. Figure 13e and f, Case History 9, shows raypaths from a dipping reflector in map view and perspective. The sideswipe is clearly shown in map view by the reflection points located off the seismic line.

If a reflector has any component of dip in the direction normal to the plane of the seismic section, termed "crossline dip", the dip will give rise to sideswipe reflections. By this definition, sideswipe would occur even where the crossline dip is slight, and so, strictly speaking, this type of event is the rule rather than the exception. (The exceptions occur where a reflector has no seismically discernible structural dip, or where the seismic line is oriented so closely to structural dip that the reflection has no seismically discernable crossline dip.)

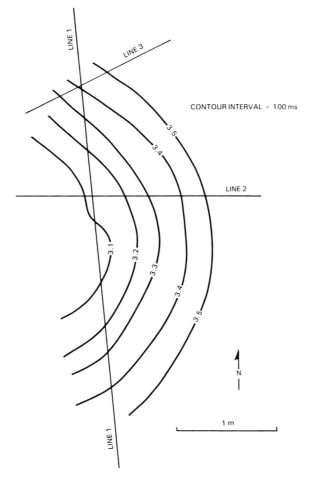

Fig. 62. A reflection map (constructed from unmigrated seismic sections). This map shows that Line 2 is in far better dip orientation than Line 1 and is likely to provide the more accurate migrated image.

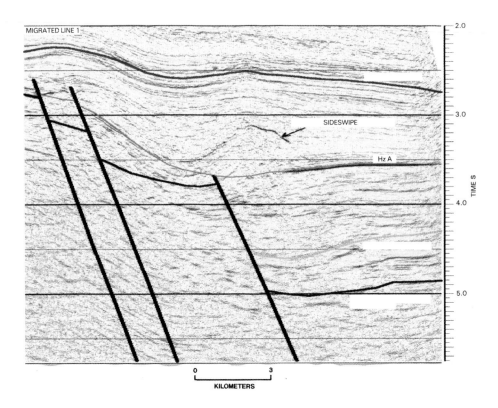

Fig. 63. An interpretation of migrated Line 1 showing the high side event as a sideswipe event.

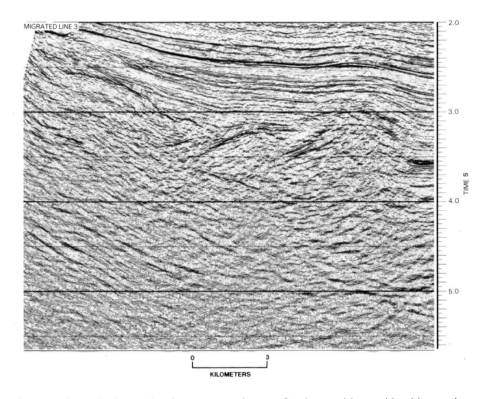

Fig. 64. Migrated Line 3 showing two prominent reflections, with considerable overlap, from Horizon A.

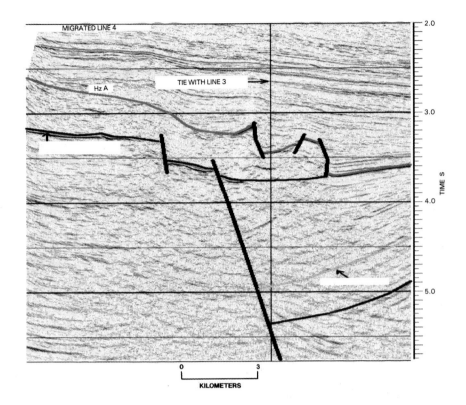

Fig. 65. Migrated Line 4 indicates only one event in the tie position.

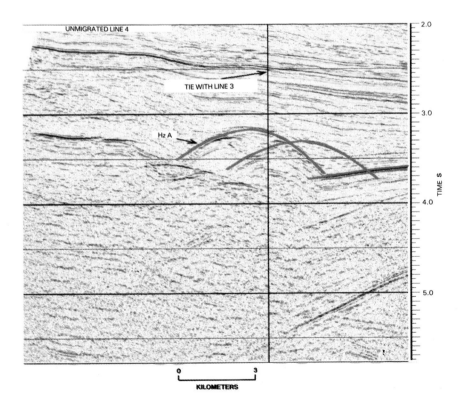

Fig. 66. Unmigrated Line 4 shows that two diffraction events, from block edges north and south of the tie position with Line 3, extend into the tie position.

If crossline dip is mild and sideswipe is slight, there will be only negligible imaging distortion. In these cases the sideswipe is probably not worthy of consideration. Therefore, in practice and in this volume, the term sideswipe is used only for those reflections where out-of-the-plane effects have resulted in significant imaging distortions.

Figure 56 shows a seismic section with an easily recognizable example of sideswipe. A prominent antiformal event crests at about 4.2 s in the middle of the section. Superimposed on this event are through-going and cross-cutting reflections dipping gently to the left. Clearly, to develop a structural interpretation that would incorporate surfaces corresponding to both these events would be difficult. The interpreter will suspect that the antiformal event is sideswipe given the knowledge that the seismic section lies along the margin of a prominent salt structure (in Case History 9 it is shown how one can precisely associate this event with a top-of-salt reflection).

The superposition of images is the most recognizable manifestation of sideswipe. Unfortunately, sideswipe occurs more frequently and in far more subtle ways than this example would suggest. Indeed, there will always be some seismic sections containing reflection sideswipe in any 2-D grid over complex structure. A seismic line aligned in a strike direction will record reflections from steep interfaces well off the line of section. If the line is sited over a structural low, there often will be sideswipe reflections from both flanks of the structure as shown in Figure 13e and f, Case History 9. These repeated reflections, if not understood, may lead to the inference that a surface is tectonically repeated and that crustal contraction has occurred.

Figures 57 to 68 present examples which illustrate these effects. In Figure 57 a

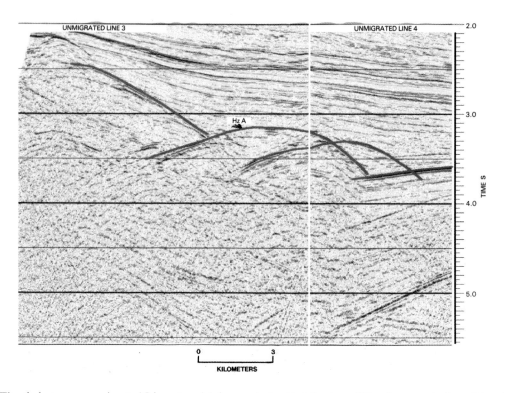

Fig. 67. The tie between unmigrated Lines 3 and 4 demonstrates that the two diffraction events from Line 4 are the same reflection as the overlapping events on Line 3. Because migrated Line 4 shows these events collapse out of the tie position with Line 3, the sideswipe nature of the events on Line 3 is proven.

migrated time-structure map depicts a salt-cored, box-shaped uplift bounded by faults on both northern and eastern margins. The southern margin is a synclinal axis about which the structure rotated during uplift. The structure plunges to the west and is segmented by several small east-west faults.

Migrated Line 1 (Figure 58) lies just to the east of the uplift. Two events from Horizon A appear to be in reverse offset as interpreted in Figure 59. However, the tie with Migrated Line 2 (Figure 60) indicates that the high-side reflection of Horizon A does not extend as far east as Line 1. Figure 61 shows that this event ties between unmigrated Lines 1 and 2. A reflection map of this unmigrated event (Figure 62) indicates that Line 2 is in good dip orientation and should provide a correct migrated image. The highside event on Line 1 should be interpreted as a sideswipe reflection from the eastern margin of the structure as shown in Figure 63.

A second, perhaps more profound, instance of sideswipe is shown over the same structure. Migrated Line 3 (Figure 64) depicts two prominent reflections from Horizon A, with considerable overlap between 3.2 and 3.5 s in the right central part of the section. However, migrated Line 4 (Figure 65) indicates only one event in the tie position. Migrated Line 4 shows that Line 3 is sited in a graben between two horst blocks. Unmigrated Line 4 (Figure 66) shows that diffraction events from these block edges extend into Line 3 and tie to the two events on that line (Figure 67). Therefore, both the events on Line 3 are sideswipe with reflection points located along block edges to the north and south of the line (Figure 68).

An important conclusion to be drawn from these examples is that sideswipe often cannot be recognized by examining only the line the sideswipe occurs on, particularly if the section is migrated. Generally, the cross line relationships on the unmigrated sections must be examined; after all, sideswipe is a 3-D effect. There is only one way to verify sideswipe and it is nearly foolproof: *if there is*

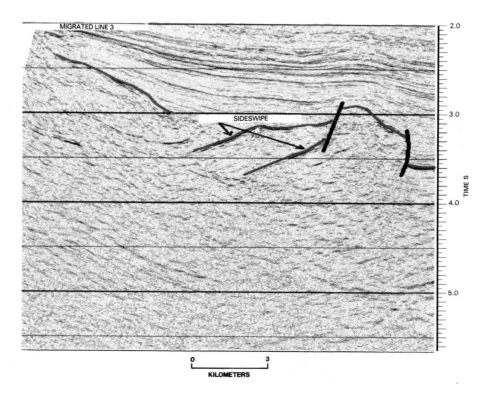

Fig. 68. Interpretation of Line 3 showing the two overlapping events as sideswipe.

crossline dip on the unmigrated tie lines, there must be sideswipe. If there is no dip (or if the dip is slight), then there almost certainly is no sideswipe.

It has been shown how sideswipe can be recognized in a grid of seismic data. This recognition is essential for properly constructing the reflection maps that are to be compared to 3-D forward modeling results or used directly in inversion. Also shown is that sideswipe can result in more than one reflection from the same surface (this of course can also arise without sideswipe). In the next section we discuss how these types of reflections are mapped.

Multivalued reflections

Recalling the simple field experiment described in the beginning of this chapter, some receivers may record more than one reflection from the reflector. This is particularly common where subsurface synclines occur which give rise to "bow-tie" reflections. The question arises as to how to map multivalued reflections.

Figure 69 shows a bowtie reflection from a syncline. The bowtie is composed of three segments coming from the left flank, right flank, and trough of the syncline. To fully define the reflection for inversion, the interpreter would outline each of these three reflection segments as shown in Figure 70. Figure 70 shows how each of these segments can be migrated (successively or concurrently) using inverse ray tracing to fully define the initial syncline.

Unfortunately, to extend this reflection definition procedure to three dimensions is time consuming and difficult. The main problem occurs where a surface is complexly folded or flexed. In these cases, the number of reflection segments could become intolerable. Moreover, if in areas up-plunge along the syncline the reflection is no longer multivalued, it would be difficult to merge the single reflection map with the multivalued one. Some shortcuts are necessary.

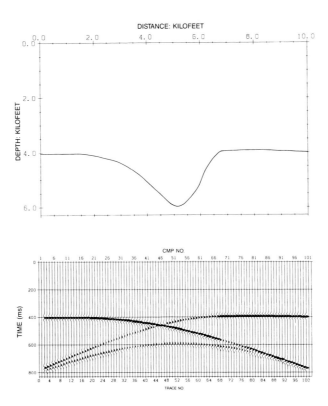

Fig. 69. A syncline and its associated bowtie reflection.

One shortcut would be to map only the earliest arrival (Figure 71) and create only one reflection map. The main drawback to this approach is that the interpreter cannot utilize the reflection information from the later arrivals, which represents the control for the trough of the structure. Figure 4.24 shows the results from this approach. Notice that the interpreter must depend on interpolation to define the trough. Notwithstanding these problems, this procedure is by far the simplest, and if resolving the trough of a particular structure is unimportant in defining the modeling objective (for example, if raypaths through the trough do not reflect off of the prospect structure), then this is the appropriate procedure.

If some attention to the trough is warranted, the most practical approach is to define two reflection segments rather than three, as shown in Figure 72. In this case, the trough is defined within a separate map. Up-plunge, where the syncline is no longer multivalued, this second map ends and there is no merge problem. As can be seen in Figure 72, although some control is lost along the middle of each

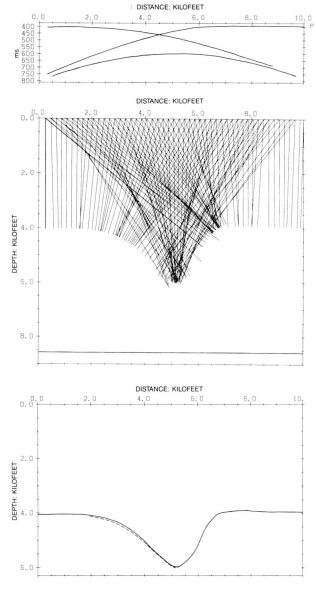

Fig. 70. Defining the bowtie by ray trace inversion of three reflection segments. Dashed line indicates the original, modeled syncline.

of the syncline flanks, the syncline is well defined. Of course, in interpreting multivalued reflections the interpreter's options are limited by the signal present on the section. If only the first arrival is apparent on the stacked section, as is often the case, then the first arrival is all that can be mapped.

MAP MIGRATION

As shown in Figure 4, zero-offset 3-D inversion is referred to as map migration. Map migration is sometimes viewed as an adjustment the interpreter makes on a structure map derived from unmigrated or time-migrated data. Although map migration can be viewed in this limited way, it is in fact one of the most powerful interpretive approaches that can be brought to the problem of complex structure interpretation. Map migration is an inversion process which transforms reflection maps into depth maps. Provided with accurate reflection maps and layer interval velocities, map migration can be used to derive a geophysically compatible depth

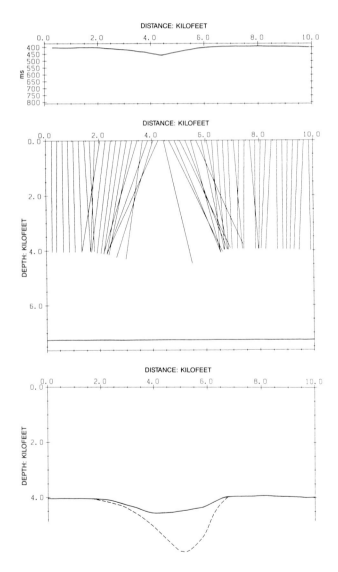

Fig. 71. Defining the bowtie by ray trace inversion of one reflection segment. Note the lack of constraint in the trough portion of the syncline. Dashed line indicates the original, modeled syncline.

solution that properly accounts for sideswipe, raypath bending, and velocity anomalies.

Concepts related to direct inversion were outlined in Chapter 2. These concepts were presented from a 2-D standpoint but can readily be generalized to three dimensions. An example of map migration operation is given in Case History 9, and the following discussion refers to Case History 9 figures. (Another map migration example is presented in Case History 10). In three dimensions, reflection interpretation results in the generation of reflection maps rather than 2-D line segments. (Case History 9, Figure 4). If reflections are multivalued then several maps may need to be generated for a single reflection as described in the previous section (Case History 9, Figures 5a, b and c).

As inverse rays are traced downward, departure angles must be determined in three dimensions such that the ray is traced parallel to the dip-azimuth of reflection structure. Rays are initiated over the map area at some grid spacing, and each is traced until reflection time is accounted for. The result is a series of depth

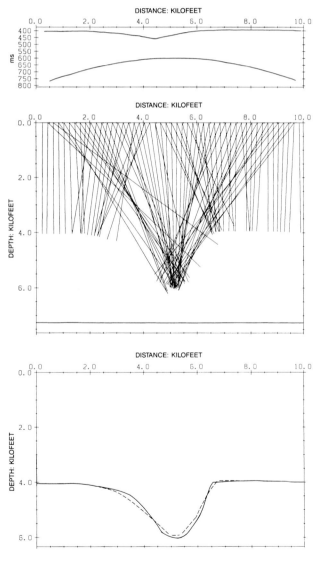

Fig. 72. Defining the bowtie by ray trace inversion of two reflection segments. Note that the syncline trough is constrained by the second smaller reflection segment. Dashed line indicates the original, modeled syncline.

points corresponding to each ray terminus (Case History 9, Figure 7). The depth points developed for multivalued reflections should not exhibit overlap (Case History 9, Figure 8). If they do, then velocities or reflection interpretation should be checked.

The depth points defined by the ray tracing are used to derive a structural model. This model will generally be geophysically compatible (i.e., capable of recreating the unmigrated section. Case History 9 shows by forward modeling that geophysical compatibility was achieved for an entire grid of eleven lines (Figures 9–19). In three dimensions, such compatibility is extremely difficult to achieve without the use of inversion. Moreover, the reflection interpretation and modeling steps give the interpreter insight into the nature of the unmigrated section and develop confidence in the interpretation. Examples of this are the modeling results of Line A and Line 4 in Case History 9. In both cases, the multiple base-of-salt events would be difficult to decipher without the aid of modeling.

As a final note, map migration of a reflection requires 3-D definition of a reflection. The reflection must be continuous enough to define on at least two lines. Sections which do not display reflections on a stacked section cannot be map migrated. Moreover, if a reflector is segmented into numerous narrow fault blocks reflector continuity will be insufficient for map migration.

CHAPTER 5

Modeling Pitfalls

Several pitfalls associated with modeling are related to either the sources of error in modeling or to misuse of modeling results.

SOURCES OF ERRORS IN MODELING

Four sources of significant error in ray-trace modeling are (1) errors in reflection interpretation, (2) errors in velocity estimation, (3) errors in sideswipe, and (4) errors related to the CMP assumption.

Errors in reflection interpretation

Errors made in reflection interpretation include, interpreting events other than primary reflections, and misidentifying a reflection (for example, mistaking a fault plane reflection for a subthrust reflection). But perhaps the most common error is to interpret too much of a reflection. If reflection continuity is poor, and the interpretation extends beyond the area where the reflection could be easily discerned, errors will usually result. This error is particularly common where one attempts to interpret through a shadow zone, perhaps caused by an overlying velocity anomaly. The common instinct in these situations is to estimate the missing reflection by what one views as the likely depth structure. However, unless one properly gauges the appropriate pull-up or sag in the reflection, interpretation error will always result. For instance, in Figure 73 the reflection is missing below the salt structure. The salt structure is regionally known to be detached from (does not involve) the underlying section. Therefore, the interpreter reasonably infers that the subsalt reflector structure is unaffected by the overlying salt structure and estimates the reflector to be planar and throughgoing. Let us see the consequence of this inference. In Figure 74a, the depth structure of the reflector below the salt has been derived by inverse ray tracing. The model includes a sag below the salt structure. Figure 74b shows the synthetic seismic section obtained by forward modeling this depth structure. A relatively flat reflection is shown at 1800 ms which is in agreement with the observed section shown in Figure 73. Although bow-tie tails associated with the syncline are present on the synthetic section and not on the observed section, the interpreter might invoke a variety of reasons why the bow-tie tails were not imaged (for example, over much of their extent they have lower amplitude). Because there are no observations on the observed section that are in disagreement with the forward model, the interpreter may well conclude that a geophysically plausible model was derived.

However, there are good reasons to consider the solution implausible. The basement structure that has been derived exactly cancels the velocity pull-up that would ordinarily be associated with the salt structure. Surely, this is too much of a coincidence. The key here is that the interpreter should only interpret the portion of reflection structure that can actually be observed and no more. Of course to estimate structure where seismic constraint is lacking is proper. But this should not be done by inferring unmigrated reflection structure when the ultimate intention is to inverse ray trace that reflection. In the example given, the interpreter should fill in the subsalt structure after the inversion of observable reflection. Interpreting reflection structure below anomalous velocity bodies requires accurate definition of the time pull-up or sag. This is extremely difficult to do if the pull-up or sag has not been imaged on the stacked section.

This sort of error occurs often in modeling and in much more subtle ways than in this example. The error is particularly common in modeling thrust sheets where there is an almost irresistible temptation to "fill in" additional reflection structure at leading-edge hanging-wall positions or at footwall truncations. The most important test that can be applied to guard against this pitfall is: *The structure in depth should not be more complex than the structure in time*. A structure can be greatly modified or made more complex in time (as a velocity pull-up in time of a reflection from an undeformed, subsalt basement is more complex than the flat basement in depth). But a fold or fault block is extremely unlikely to have an entirely unstructured appearance in time because of velocity effects. For this circumstance to occur requires the fortuitous arrangement of velocity bodies. This rule should be adhered to in any time/depth conversion procedure which, broadly considered, is what structural modeling is.

Velocity errors

Any errors in estimating layer velocities translate directly into errors in depth structure estimation. In fact, one can recognize a level of ambiguity in the structural solution that corresponds to the uncertainty in the velocity estimates.

Figure 75 shows a variety of subsurface models (shown at left) each with a different structure and different layer velocities. Each model has a seismic response (shown at right) that is virtually identical. Thicker layers have faster velocities, and so traveltimes through these layers are uniform. The figure demonstrates that the uncertainty in the depth solution is related to the extent to which subsurface velocities are known. It is often useful to assess this uncertainty by modeling a range of cases which represent the range of plausible layer velocities.

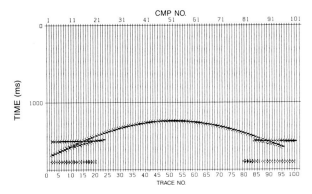

Fig. 73. A stacked, unmigrated, seismic section over a salt dome. The subsalt reflection is absent below the dome, but the interpreter has inferred the reflection to be flat because regionally the subsalt section is known to be undeformed.

A related error arises in the way velocity variation is represented. Velocities always vary laterally and vertically. As shown in Figure 38, if vertical gradients are great, then the resulting raypath will be highly curved. If the modeling software does not directly simulate curved raypaths, an error will result. In these cases, the error may be minimized by utilizing several layers, of constant velocity, to represent the vertical gradient.

Sideswipe

When modeling in two-dimensions, there is an assumption that all of the raypaths are contained within the plane of the section. To the extent that this is not true, there will always be some error in the 2-D solution. As was described in Chapter 4, the magnitude of the sideswipe problem is best gauged by examining the crossing seismic lines for crossline dip.

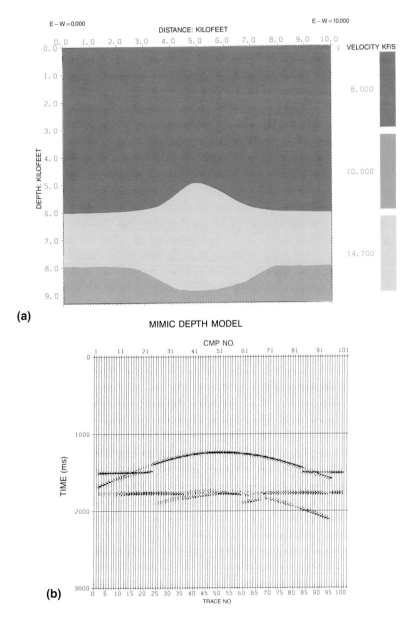

Fig. 74. (a) The structural model derived by inverse ray tracing. Note that a structural low has been derived in the subsalt section. (b) The arrival time response replicates the visible reflections in Fig. 73.

The usual consequence of modeling a sideswipe reflection is to erroneously estimate the depth of a reflection. If the sideswipe is relatively minor, the depth is underestimated. This effect was implied in Figure 55 where the reflection along Line A, the correctly migrated dipline, is deeper (at the line intersection) than the reflection is along Line B where the reflection is sideswiped.

Alternatively, the sideswipe may be extensive; for example where a feature is

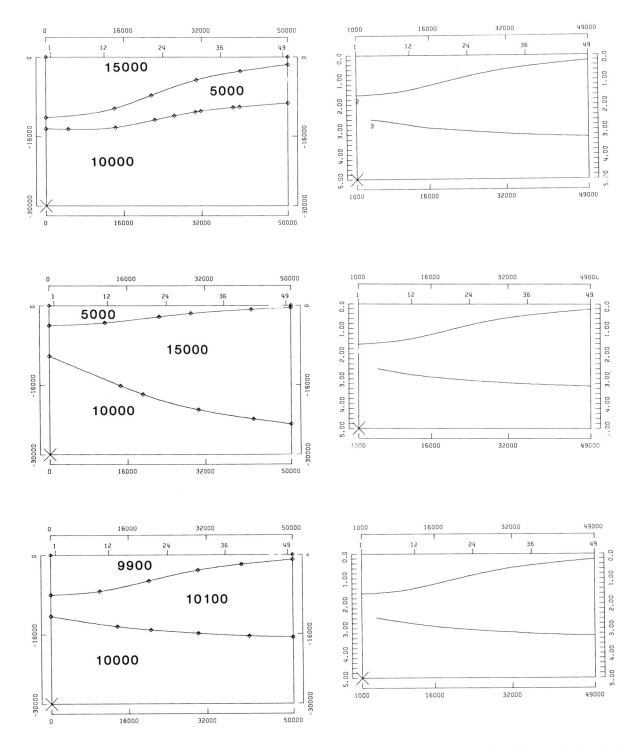

Fig. 75. A series of models and their arrival time responses. Although the models are quite different, their arrival times are identical.

recorded along a seismic line which does not include the feature in its vertical plane. In these cases, the added raypath length from the horizontal component causes greatly increased traveltime and may result in the feature appearing too deep in the section. A relatively surficial feature may appear at basement level. An example of this type is shown on the seismic line in Case History 9 where the sideswiped top of salt event appears about a second deeper than the neighboring lines with little or no sideswipe (compare Lines 1 and 2 from Case History 9).

As described in Chapters 2 and 4, in modeling the stacked section, it is assumed that the section can be represented by a zero-offset, normal-incidence raypath. This source of error is the most difficult to assess but, as was indicated in Figure 54, the error is almost certainly quite a bit less than the greatest moveout value in the gather. In the situation diagrammed in Figure 54, the stacked section would depict an earlier reflection arrival time than the actual T_0 at the gather location. Modeling would yield a structure that was too shallow.

In addition to these structural errors, the inferences one might draw about acquisition conditions from zero-offset modeling may be invalid. For instance, the shadow zone which was shown in Figure 40 may exist only for zero-offset shooting, while the farther offset traces may avoid the critical angle effects by undershooting the structure.

As stated, these are the only sources of error in modeling. If a reflection map is well simulated, then challenges to the derived structure should be based on one or more of these errors. In addition, one may always question those aspects of a depth map that are not constrained by modeling. These include those portions of the depth map which do not contain reflection points associated with observed seismic reflections.

MISUSES OF MODELING

There are several misuses of modeling which are worth mentioning. One has to do with using ray tracing results as a seismic facies. The interpreter might mistakenly feel he has successfully modeled his data when he has merely recreated a chaotic or disorganized seismic facies. In reality, because so many complex structures yield chaotic facies patterns, he has proved little by this exercise.

A second misuse involves the presentation of raypath results. Often a highly complex raypath diagram is used to make the point that the imaging of the structure would be a complex task. In other words, the diagram can be used to suggest that seismic data cannot be used to define the structure. However, one should bear in mind that even highly complex structures can be successfully imaged by depth migration programs, when synthetically modeled (see the examples of depth migration of synthetic data shown in Case History 4). Admittedly, synthetic data sets are noiseless and the subsurface velocity structure is completely known. But complex raypaths alone do not demonstrate that a structure cannot be imaged.

Finally, in modeling there is a temptation to expend effort simulating aspects of the seismic data that do not constrain structure. For example, adding random noise to data will make the simulated section look more like the real one, but offers no additional insight into the structure modeled. Convolving the arrival times with a wavelet usually falls into the same category. The interpreter should only simulate those aspects of the data which offer some prospect of influencing the structural solution.

REFERENCES

Bregman, N. D., Bailey R. C., and Chapman C. H., 1989, Crosshole seismic tomography: Geophysics, **54**, 200–215.

Bishop, T. N., Bube, K. P., Cutler, R. T., Langan, R. T., Love, P. L., Resnick, J. R., Shuey, R. T., Spindler, D. A., and Wyld, H. W., 1985, Tomographic determination of velocity and depth in laterally varying media: Geophysics, **50**, 903–923.

Canal, P., and Diet, J. P., 1987 Interactive interpretation and 3D modeling: 12th World Petrol Cong., Houston, Proc. **2**, 119–124.

Chiu, S. K. L., Kanasewich, E. R., and Phadke, S., 1986, Three-dimensional determination of structure and velocity by seismic tomography: Geophysics, **51**, 1559–1571.

Cutler, R. T., 1987, Seismic tomography: Presented at the 57th Ann. Internat. Mtg., Soc. Expl. Geophys.

Dix, C. H., 1955, Seismic velocities from surface measurements: Geophysics, **20**, 68–86.

Gardner, G. H. F., 1985, Migration of seismic data: Geophysics reprint series, No. 4; Soc. Expl. Geophys.

Gerritsma, P. H. A., 1977, Time-to-depth conversion in the presence of structure: Geophysics, **42**, 760–772.

Hardage, B. A. 1983, Vertical seismic profiling: Geophysical Press, London-Amsterdam.

Hron, F., and May, B. T., 1978, Synthetic seismic sections of typical petroleum traps: Geophysics, **43**, 1119–1147.

Hubral, P., 1977, Time migration—some ray-theoretical aspects: Geophys. Prosp., **25**, 738–745.

Kennett, B. L. N., and Harding, A. J., 1985, Is ray theory adequate for reflection seismic modelling? (a survey of modeling methods): First Break, **3**, 9–14.

Larner, K., Hatton, L., Gibson, B., and Hsu, I., 1981, Depth migration of imaged time sections: Geophysics, **46**, 734–750.

Levin, F. K., 1971, Apparent velocity from dipping interface reflections: Geophysics, **36**, 510–516.

May, B. T., and Covey, J. D., 1981, An inverse ray method for computing geoloic structures from seismic reflections—zero-offset case: Geophysics, **46**, 268–287.

May, B. T., and Covey, J. D., 1983, Structural inversion of salt dome flanks: Geophysics, **48**, 1039–1050.

Miller, M. K., 1974, Stacking of reflection from complex structures: Geophysics, 39, 427–440.

Neidell, N. S., and Poggiagliomi, E., 1977, Stratigraphic modeling and interpretation—geophysical principles and techniques., in Payton, C., Ed., Seismic stratigraphy—applications to hydrocarbon exploration: Am. Assn. Petr. Geol. Memoir 26.

Pereyra, V., 1988, Two-point ray tracing in complex 3-D media: Presented at the 58th. Ann. Internat. Mtg., Soc. Expl. Geophys.

Sarvesam, G., and Dhole, A., 1986, Image ray tracing—a method for seismic modeling: J. Assoc. Expl. Geophys. (India), **7**, 77–83.

Shah, P. M., 1973, Ray tracing in three dimensions: Geophysics, **38**, 600–604.

Skeen, R. C., and Ray, R. R., 1983, Seismic models and interpretation of the Caspar Arch Thrust: Application to Rocky Mountain foreland structure: in Lowell, J. D., Ed., Rocky Mountain foreland basins and uplifts; Rocky Mountain Assn. Geologists, 99–124.

Taner, M. T., Cook, E. E., and Neidell, N. S., 1970, Limitations of the seismic reflection methods—Lessons from computer simulations: Geophysics, **35**, 551–573.

Thorn, S. A., and Jones, T. P., 1986, Application of image ray tracing in the Southern North Sea gas fields: Habitat of Paleozoic gas in NW Europe, Conf. Geol. Soc. London Spec Publication 23, 169–186.

Tucker, P., and Yorston, H., 1973, Pitfalls in seismic interpretation: Soc. Expl. Geophys. Monograph 2.

Waltham, D. A. 1988, Two-point ray tracing using Fermat's principle: Geophysics, **93**, 575–582.

Wason, C. B., Black, J. L., and King, G. A., 1984, Seismic modeling and inversion: Proc. Inst. Elect. Electron. Eng., **72**, 1385–1393.

Withjack, M. O., Kristian, E. M., and Reinke-Walter, J., 1987, Seismic expression of structural styles: A modeling approach: Presented at the Ann. Am. Assn. Petr. Geol.—Soc. Econ. Paleont. Mineral. Mtg., **71**, No 5, 628.

Yancey, M. S., and McClellan, B. D., 1983, Drape fold, central Wyoming: in Bally, A. W., Ed., Seismic expression of structural styles, Vol. 3, Am. Assn. Petr. Geol. Studies in Geology Series 15.

Yilmaz, O., 1987, Seismic data processing, Investigations in Geophysics Series 2; Soc. Expl. Geophys.

PART II

Modeling Case Histories

INTRODUCTION

Each Case History presented in Part II exemplifies a different aspect of seismic modeling. In each one insight was gained into some aspect of the structure under investigation. The examples are almost entirely from either thrust belts or zones of salt mobilization. The examples reflect not only the dominance of these two environments in complex-structure exploration, but the tendency for raypath bending and velocity effects in these structures.

Case History 1, by Winkelman and Hall, illustrates basic operation of forward modeling. Unmigrated reflection arrival times from beneath a salt sill are simulated by normal incidence ray-tracing. The derived structure looks very different from the seismic section because velocity and raypath effects have been removed. The final solution is used to define the velocity field for depth migration which results in a depth image identical to the ray-traced solution.

Case History 2, by McClellan et al., is actually a series of case histories. Each demonstrates a different time/depth conversion technique appropriate for a different level of structural complexity. In this way the case histories provide a context for demonstrating the role of seismic modeling within the spectrum of time/depth conversion problems faced in the exploration environment.

Case History 3, by Lingrey, demonstrates the interaction of seismic modeling with an allied technique, geometric modeling of complex structures. When the two modeling techniques are applied in combination the depth solution can be constrained far more rigorously than when either is applied alone. This case history also demonstrates the usefulness of the basement velocity anomaly as a modeling parameter which constrains the total thickness of imbricated, high-velocity section.

Case History 4, by Morse et al., also applies geometric modeling in combination with seismic modeling in investigating an imbricated structure, but with some differences. The geometric modeling is applied in a forward direction (with respect to time) rather than as a restoration. Also, the imbricated structure is subsiding and exhibits section growth on both the leading edge and trailing edge flanks. As a consequence, subsurface velocities are a function of present day depth of burial rather than age, as is more typical in thrust belts. The 3-D geometric models created allow the authors to explore 3-D imaging effects. Finally, wave equation modeling is used to create synthetic data sets which are used to investigate imaging strategies.

In Case History 5, by Nosal et al., seismic modeling is used to design a salt proximity survey with the objective of defining the flank of a salt dome. Specifically, modeling was used to define the portion of the salt flank that would be illuminated by a potential shot location. The modeling was also used to aid the interpretation of the data that was ultimately collected. Subsequent well results agreed well with the modeling prediction.

Case History 6, by Zijderveld et al., demonstrates the use of stacking velocity values as a modeling parameter. Although stacking velocities do not resemble rms velocities in complexly structured environments, they are a predictable consequence of subsurface velocities and structure. In the technique presented a model is automatically perturbed until modeled arrival times and stacking velocities agree with observed ones. By engaging in the extra effort of modeling stacking velocities rather than just arrival times, layer interval velocities can be derived rather than assumed.

Case History 7, by Zimmerman, demonstrates the use of physical modeling for investigating structures. The techniques and apparatus relating to physical modeling are described and contrasted with ray-trace modeling. A simulation of a seismic line over a salt sill is presented.

The next three case histories demonstrate 3-D modeling. In Case History 8 Rudolph and Greenlee discuss the use of ray-trace modeling to define the flank of a reef. The principal seismic manifestations of the reef are diffraction limbs emanating from the reef crest. The 3-D modeling shows that lines which are positioned off the reef crest, and even away from the reef, will yield a highly distorted picture of the reef because of sideswipe.

Case History 9, by Fagin, demonstrates the use of map migration and 3-D forward modeling to define the base of a salt sill. By employing an inversion technique a model was derived which would closely simulate eleven lines in a seismic grid when ray traced. Most of the lines depict a strongly sideswiped base of salt reflection that results in a highly misleading structural image on migrated sections.

Case History 10, by Reilly, also presents a map migration case history from the North Sea. Two vendor software programs are analyzed, one which simulates curved raypaths and one which does not. In addition, a VSP survey is modeled to determine whether arrivals support the map migration solution.

CASE HISTORY 1

Seismic Modeling Beneath a Salt Flow

Benjamin E. Winkelman* and Craig E. Hall*

ABSTRACT

A high relief structure was noted on seismic data beneath a salt overhang. The prospective nature of this subsalt feature hinged upon the ability to discriminate between a structural anticline in depth and the illusion of an anticline in time. Imaging distortion brought about by: (1) velocity pull up due to the high velocity of salt; and (2) raypath bending associated with salt boundaries, complicates structural interpretation. A depth model was created from an interpreted time-migrated line to confirm the presence of a true anticline in depth. The resulting two-dimensional (2-D) synthetic seismic section compared well with the original seismic line. Finally, depth migration was performed on the original seismic data, with a result similar to that of the modeling. This result demonstrated the use of a simple modeling technique to validate a prospective exploration target.

INTRODUCTION

Salt domes were the first offshore exploration targets in the Gulf of Mexico. Initially, gravity maps and low-fold seismic data provided the basic information to locate exploratory wells over these features. With the advent of more powerful seismic sources, longer hydrophone arrays, and improved processing techniques, more detailed interpretations of salt structures have become possible. The Sigsbee Escarpment salt overhang is an excellent example of where these acquisition and processing techniques have been employed to improve imaging. Originally interpreted as a salt massif, recent seismic data reveal events in what was thought to be the heart of the massif. This data has led to the reinterpretation of the Sigsbee Escarpment as a salt sill. (Amery, 1969).

Even with the improved imaging, many interpretational questions nevertheless remain, particularly in evaluating subsalt structure. For example, high-velocity salt is notorious for producing "pull up" of underlying structures. In addition, time migration may result in significant lateral mispositioning of events, particularly when migrating reflections beneath the surface produce strong raypath bending such as the top and base of salt. Depth migration may be employed to properly account for these effects.

With these considerations in mind, we report on our investigation of a high-relief time structure underlying a salt flow in an offshore Louisiana area

*ARCO Oil and Gas Company, Houston, Texas.

which has led to the definition of a new exploration target. Modeling was employed to test the plausibility of the initial interpretation and to guide the depth migration used in the final imaging.

METHOD AND RESULTS

An initial time map was constructed from anticlinal events picked on several seismic lines and tied at line intersections. The events occurred on several vintages of data, though the most recent data imaged them best. The possibility that the anticlinal events were due to multiples was ruled out because rms velocities for the reflections were too high.

We modeled a line in this area in an attempt to verify the structural interpretation. Sonic logs from nearby wells were used to determine salt and sedimentary layer velocities. The sonic log data indicated that the salt velocity is roughly twice as fast as that of the surrounding sedimentary layers. Interval velocity within each layer was held constant laterally through the model. This was felt to be a reasonable approximation based upon the control available. Figure 1 shows the

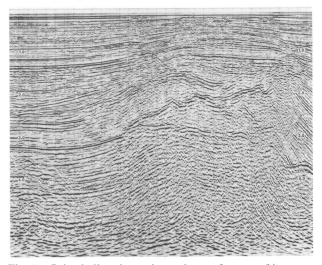

Fig. 1a. Seismic line time-migrated over feature of interest.

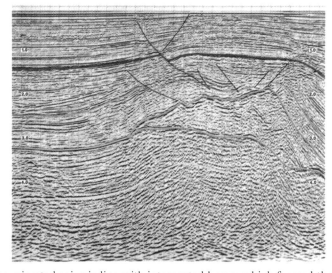

Fig. 1b. Time-migrated seismic line with interpreted layers which formed the basis for the modeling. The salt is outlined in orange. The objective horizon is yellow.

migrated seismic line over the feature of interest and the interpreted layers which formed the basis for the modeling. The line was shot using a conventional air-gun array and streamer cable. Because the line is oriented perpendicular to the front of the salt flow, we thought 2-D modeling would be appropriate. Interpreted reflections were chosen by amplitude and velocity trend. Figure 2 shows the time model.

This interpretation in time was then converted to depth. Figure 3 shows the model in depth. Note the difference in the appearance of the anticline displayed in depth versus time. The use of displays of this type allowed us to compare the resulting model to a geologic cross-section. With the imaging distortions due to the velocity contrasts removed, the task of judging the relative soundness of the geologic interpretation made from the seismic data was greatly simplified. At this stage, various cases testing alternative seismic interpretations of the thickness of the salt layer were run. Persistence of the anticline in depth under these possible interpretation variations confirmed the prospective nature of the feature.

Normal incidence ray tracing was then performed on the depth model. This exercise served as a check on the time-migrated section from which the depth model was constructed. An appropriate wavelet was determined from phase and frequency analysis of the seismic data relative to the well control. The wavelet was then convolved with the spike section resulting from the ray-trace modeling

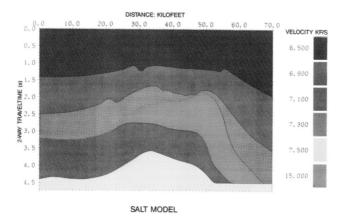

Fig. 2. Input time model from Figure 1b. Roughness in the shallow horizons is due to minor faulting.

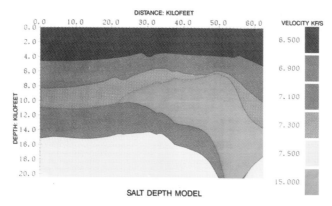

Fig. 3. The model in depth resulting from the conversion of the time model, using velocity control from nearby wells and the integration of seismic derived velocities. Note the anticlinal expression of the yellow horizon in depth (color scale as in Figure 2).

to produce Figure 4. This section was judged to be a satisfactory simulation of the original seismic line. This result confirmed the plausibility of the original interpretation. Comparison with the model in depth dramatizes the differences between imaging in the time domain and in the geologic structure.

The asymmetric nature of the salt flow brought into question the proper lateral positioning of the subsalt structural crest. The relative simplicity of the time-to-depth conversion is attractive from an economic point of view, but only works well in areas where lateral velocity changes aren't too complex. As a further check on the interpretation, depth migration was performed on the seismic line. Figure 5 shows the resulting section. This section compares quite favorably to the model in depth shown in Figure 3. Only a very slight lateral shift in subsalt structural position is evident, thus further confirming the modeling results.

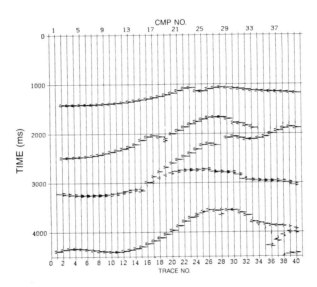

Fig. 4a. The synthetic section resulting from normal incidence ray tracing performed on the salt depth model.

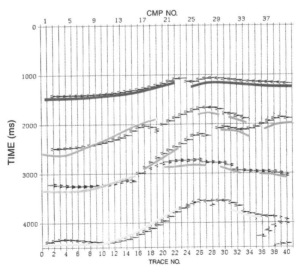

Fig. 4b. The unmigrated seismic interpretation superimposed on the synthetic section.

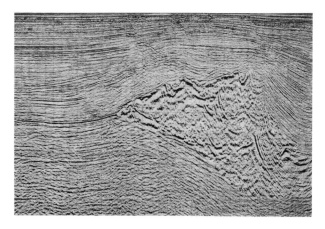

Fig. 5a. Uninterpreted, depth-migrated, seismic section.

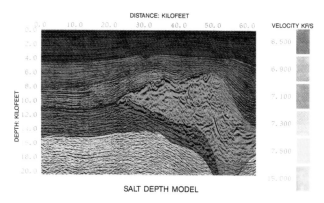

Fig. 5b. The depth model from Figure 3 superimposed on the depth-migrated seismic section. Note the anticlinal expression of the yellow horizon on both displays.

SUMMARY

Mapping and analysis of the structural style of the surrounding area lent geologic credence to the presence of a structural high. This credence, combined with the time-to-depth conversion showing that the anticline was not merely a velocity artifact, led to serious pursuit of the feature as an exploration target. The problem then became one of feature definition. How much structural depth relief is present, as opposed to seismic time relief? After accounting for the salt velocity in the overhang, how deep is the prospect? Answers to these questions were readily apparent from the modeling results. The time map was converted to depth using the model's velocity function. The resulting map showed a large structural closure under the salt overhang. Drilling depth to the anticline was shown to be quite feasible. In summary, the modeling process allowed the limited geologic control to be used with the seismic data to characterize the prospect with a high degree of confidence.

REFERENCES

Amery, G. B., 1969, Structure of Sigsbee Scarp, Gulf of Mexico: Bull. Am. Assn. Petr. Geol., **58**, 2480–2482.

CASE HISTORY 2

Seismic Depth Conversion and Migration—Techniques and Applications

Bruce McClellan*, Jim Johnson*, Steve Whitney*,
and Robert Bryan*

ABSTRACT

A spectrum of time/depth conversion techniques are presented which, in order of presentation, require greater effort as needed for increasingly complex structure. These techniques are (1) using regionally smoothed stacking velocities, (2) using horizon-keyed interval velocities, (3) using iterative forward ray-trace modeling, and (4) using modeling in conjunction with depth migration. Case histories are presented which illustrate the operation and effect of each of these techniques.

INTRODUCTION

Seismic depth conversion and depth migration techniques are valuable prospecting tools in the presence of significant lateral velocity changes. Depending upon the severity of these changes, and/or the subtlety of structures, depth correction can make or break prospects.

There is an entire spectrum of lateral velocity contrasts, and the optimum method for correction may differ with severity, budget, and deadline. We discuss a few of the many available techniques and use real data examples to demonstrate them.

The four methods of seismic depth correction discussed are:

(1) Depth conversion using regionally smoothed stacking velocities,
(2) Depth conversion using horizon-keyed interval velocities,
(3) Depth conversion after seismic modeling, and
(4) Depth migration after seismic modeling.

Image-ray depth conversion, an intermediate technique that approximates the effects of depth migration, is acknowledged but not included in this review.

Presented at the 58th Annual International Meeting, of the Society of Exploration Geophysicists, October 31, 1988.
*Conoco Inc.

METHODS AND APPLICATIONS

Depth conversion using regionally smoothed stacking velocities

The simplest approach to depth conversion is using stacking velocities. Stacking velocities are an approximation of root mean square (rms) velocities, and are thus a measure of average tendency of subsurface velocity. By assuming stacking velocities are rms velocities, we can apply the Dix formula to derive interval velocities and convert to depth. There has been extensive discussion over the years regarding the validity of this assumption, and its related pitfalls. In general, the assumption works best when data quality is good, reflectors are nearly horizontal, and lateral velocity variations are gentle.

Figure 1 shows an example where stacking velocities helped to delineate a subtle prospect. At the top of the figure is a migrated time section showing a low relief structure at 1.8 s, just above the horizontal red line at 2.0 s. The lateral extent of two-way closure is yellow, with the apex of the structure indicated by an arrow.

Below the migrated section is a plot of stacking velocities to 2.0 s. There is a mild lateral velocity decrease from left to right, believed to be caused by overlying gas-charged sediments. Gas-charged sediments, like any low-velocity zone, can cause a time sag in the underlying reflectors. Sometimes these sags are obvious to interpreters, but other times they are subtle and can be missed. In this case, stacking velocities helped to both identify and correct for the sag.

Figure 2 shows the vertical depth conversion of this line, along with the original migrated time section. A smoothing operator was applied to the stacking velocities for depth conversion. Note that, in contrast to the subsequent technique, smoothing and interpolation were applied globally over the entire section and not within geologic units. On the depth section, the horizontal red reference line is the 2300 m depth line. In depth, the structure has been raised slightly on the right, enough to shift the center of the structure almost 1 km.

The example demonstrates that even very minor velocity gradients can impact exploration for subtle structures. This particular example might also serve as an end member, on the mild end, for a "spectrum" of lateral velocity-gradient image distortion problems.

As mentioned, there *are* many pitfalls to watch for when using stacking velocities for depth conversion. Problems arise when stacking velocities no longer

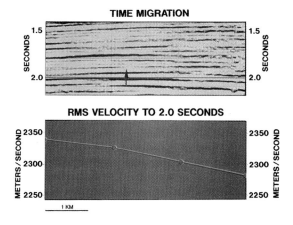

Fig. 1. At the top, a low-relief structure is shown on a migrated time section, highlighted in yellow. An arrow points to the apex of the structure. Below, the stacking velocities to 2.0 s (horizontal red line on top section) are shown. Even minor lateral velocity gradients can be important when defining gentle structures.

approximate rms velocities because of effects related to statics conditions, dip, sideswipe, raypath bending, low fold, or generally poor data quality. For example, a near-surface, low-velocity zone can cause artificially high stacking velocities due to a time sag on the near traces in a CDP gather. Each dataset must be carefully examined to evaluate, and where possible, to eliminate erroneous velocity information.

Depth conversion using horizon-keyed interval velocities

As the magnitude of lateral velocity gradient increases, so does the amount of image distortion of the structure depicted on time sections. One way to improve our description of subsurface velocities is to define velocity intervals which are associated with geologic intervals and are bounded by geologic surfaces. These are termed horizon-keyed interval velocities, and they can provide additional detail to improve depth conversions. The structure of these horizons is inferred from seismic reflectors and so interpreter involvement is an important part of the sequence.

Figure 3 is a migrated time section showing a salt layer in the deep water Gulf of Mexico. The salt layer is shown in orange. The subsalt reflector below the salt layer is indicated by a red line and is dipping gently to the left, toward a pronounced structural sag.

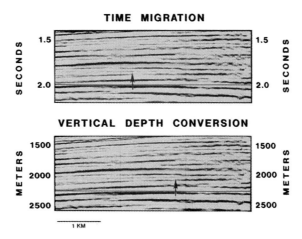

Fig. 2. The time migration from Figure 1 (above), and a vertical depth conversion obtained from the stacking velocities (below). On the depth section, the horizontal red reference line is the 2300 m depth line. The arrows show that, in depth, the center of the structure has shifted by 1 km.

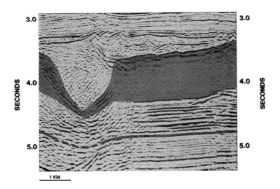

Fig. 3. Migrated time section showing a salt layer (orange) and a subsalt reflector (red). Notice the structural sag below the salt, on left.

Despite the time migration which has been applied to the data in Figure 3, the image falls well short of an adequate representation of the depth structure. Two shortcomings of time migration hinder adequate structural imaging. First, the time migration procedure cannot laterally position reflections correctly when strong lateral velocity contrasts are present. This problem is addressed in a later section on depth migration. Second, because the output of time migration is scaled in time, velocity sags and pullups remain. We seek to eliminate this second problem by employing a horizon-keyed time/depth conversion procedure. The method is shown to provide significant improvement in the structural definition of subsalt reflectors.

In this example, the top and base of the salt can be defined using velocity analyses because the interval velocity of the salt, 4400 m/s, is more than twice the 1980 m/s velocity of the overlying sediments. Lateral changes in the thickness of the high velocity salt have a strong influence on the geometry of underlying seismic reflectors. Figure 4 shows the vertical depth conversion obtained using a smoothed interval velocity distribution. The subsalt reflectors are no longer dipping to the left, and the time sag beneath the thin portion of the salt has been removed.

Figure 4 shows that reflections beneath the thin salt are poorly imaged, due to a breakdown in the assumptions behind common depth point (CDP) stacking. The problems with CDP stacking in this environment are demonstrated in Figures 5 and 6. In Figure 5, a raypath plot for a common midpoint (CMP) gather over a flat section of the salt layer indicates that there should not be a problem stacking reflections from the base of salt and subsalt reflectors. All source-receiver pairs in the gather share the same reflection points. Figure 6, a raypath plot for a CDP gather over a complex portion of the layer, shows that irregularities in the high velocity salt can cause considerable variation in reflection points, and prohibit effective stacking of the data beneath the layer. Advanced techniques, such as prestack depth migration, provide a mechanism for imaging when raypaths are highly distorted. Ray bending and CDP smear are accounted for *before* the data are stacked together, allowing the reconstruction of CDP gathers.

Depth conversion after seismic modeling

To obtain an accurate depth correction in structurally complex areas, only one thing is needed—the answer. There are different ways around this fundamental problem, but generally the solution involves trying several answers and interpreting the best one. (To paraphrase a well-known quotation, it takes one per cent inspiration and ninety-nine per cent iteration!).

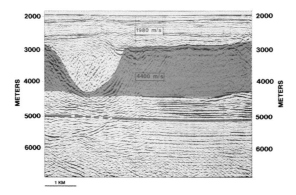

Fig. 4. Vertical depth conversion of the section in Figure 3, obtained using smoothed, horizon-keyed interval velocities. The time sag has been removed, but imaging is poor beneath the thin salt.

One technique, iterative structural modeling, allows an interpreter to quickly determine a plausible subsurface structure which results in the development of a more accurate depth velocity model ("the answer") that can be employed in both depth conversion and depth migration. The technique, an aid for both interpretation and processing, allows a quick comparison between seismic events and an interpreted subsurface, and allows the interpreter to develop a structural model that does not rest primarily on intuition.

Figures 7 through 10 illustrate the modeling procedure. In examining Figure 7, several questions arise regarding the nature of the geologic structure represented by this unmigrated section. Is the structure faulted or only folded? Is the circled structure real, or a velocity pullup? These are some of the questions that can be addressed using modeling.

Figure 8 shows the first ray-trace modeling iteration. On the left is the initial estimate of a depth velocity model, an unfaulted drapefold.

Experience shows that to start with a simple model is best, and then let the modeling procedure guide you toward more sophisticated solutions. The interval velocities for the depth model were obtained from nearby well control.

The synthetic time overlay produced from the first depth model is shown on the right side of Figure 8. At this point, only the top red horizon is being modeled. Because any change in the top event affects all the events below, we model from the top down. Visual inspection of the first synthetic time event shows that, to match the real data, we must lower the structure on the extreme left, and make a gentler forelimb on the right.

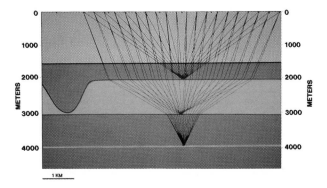

Fig. 5. Raypath plot for a CDP gather over a flat portion of the salt layer. The model shows that we should be able to stack in the base of salt and subsalt reflectors in this region. All source-receiver pairs in the gather share the same reflection points.

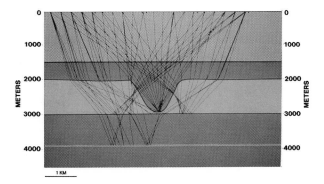

Fig. 6. CDP raypath plot over irregular salt. The complex raypath pattern and associated dispersion of reflection points results in a breakdown in the assumptions behind CDP stacking, leading to poor data quality.

After these changes are made on the depth model (Figure 9) a good match is obtained for the first layer, and we may move on to the deeper layers. Figure 9 shows intermediate progress, where the yellow and blue events are being modeled, but have not yet been matched satisfactorily. The synthetic time events fit well over much of the section, but the deeper yellow and blue events do not yet extend into the circle from Figure 7. A better result can be obtained using a high-angle reverse fault in the depth model.

Figure 10 highlights the final depth velocity model obtained from modeling. A good time match for all events is obtained using the high angle reverse fault model. While not a unique solution, the modeling provides a seismically compatible, and therefore more plausible, depth interpretation.

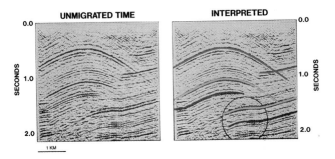

Fig. 7. Unmigrated seismic line over a drapefold structure in Wyoming, shown uninterpreted (left) and interpreted (right). Is the structure faulted or only folded? Is the circled structure real, or a velocity pullup?

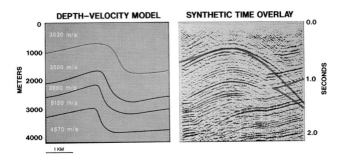

Fig. 8. Simple depth model interpretation (left), with each layer representing an interval velocity. On the right is a synthetic time plot from the depth model (first layer only) overlain on the real, unmigrated data.

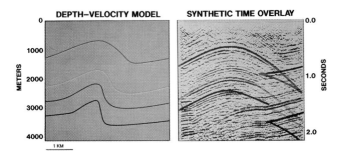

Fig. 9. Subsequent depth model and time comparison, leading to a good match for the first layer, shown in red. To match the deeper yellow and blue events in the circle from Figure 7, we introduce faulting into the depth model.

Another example of depth conversion using seismic modeling is illustrated in Figures 11 through 15. Modeling was used to produce the depth velocity model for this seismic line, and the model was used to guide seismic reprocessing and depth conversion.

Figure 11 shows an unmigrated seismic line acquired over a major thrust fault in Wyoming. To preserve steeply dipping reflectors and diffraction limbs, this line was shot using a point source (dynamite in 50 ft shotholes) and short receiver arrays.

Figure 12 shows the synthetic time match produced using iterative structural modeling. The red, yellow, and blue package of reflectors has been thrust over itself from left to right, and in front of the thrust is a pronounced "bow-tie" event, shown in brown. In the center of the subthrust section is a large anticlinal structure. Is the structure real?

The final depth velocity model, obtained by modeling, is shown in Figure 13. By comparing this depth model with its time response in Figure 12, we see that in depth there is no anticline in the subthrust section. Although the structure flattens somewhat, there is no dip to the right. The time structure is primarily a velocity pullup. The pullup is due to the strong lateral velocity contrast between units within the thrust plate and the low-velocity units in the adjacent footwall.

Traveltime modeling of key horizons is a valuable way of evaluating depth

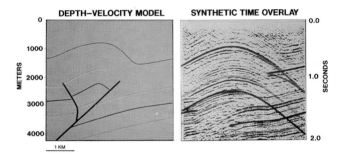

Fig. 10. The final depth velocity model and its associated time plot. A good match for all events is obtained using a high angle reverse fault model. Modeling helped to interpret this line so that the resulting depth interpretation is *compatible* with the seismic data.

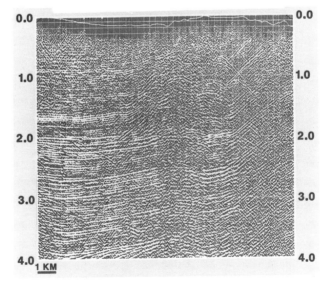

Fig. 11. Unmigrated seismic line acquired over a thrust belt in Wyoming.

interpretations. A complementary approach is to use the depth model to generate synthetic *traces*, and to compare those traces with the original data. Figure 14 shows a set of synthetic traces generated from the velocity model using a "reverse" depth migration, or forward modeling, program. The synthetic traces are a good match with the real data in Figure 11, providing additional confidence in using the model for depth conversion.

The final step in the process is depth conversion, shown in Figure 15. Model-guided processing, including vertical depth conversion, has done a good job of resolving the depth picture. The seismic traces have been "put back together" in a reasonable fashion, forming a depth image that is compatible with, and less complicated than, the unmigrated time data. Note that in conformity with the modeling result the anticline in the subthrust section is no longer apparent.

The next (and final) example illustrates how depth migration can be used to account for laterally changing velocities. In theory, depth migration is the preferred method for handling these problems. In practice, its use is somewhat

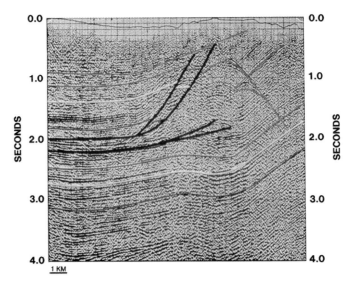

Fig. 12. The synthetic time overlay produced from the depth model, Figure 13. The red, yellow, and blue package of reflectors has been thrust over itself from left to right. The large anticlinal structure at the center of the subthrust package below 2.0 s is an attractive target . . . or is it?

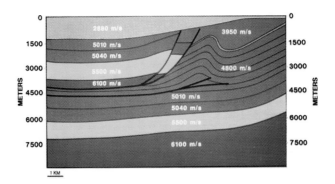

Fig. 13. The final depth velocity model, derived by iterative modeling. Older, high-velocity material in the hanging wall has been thrust up against younger, low-velocity material in the footwall, causing a subthrust velocity pullup. The subthrust "anticline" is not a real structure!

limited by greater expense and turnaround time. Advancing technology should ultimately make depth migration a standard process.

Depth migration after seismic modeling

Conventional migration normally uses some form of rms velocities to produce a generalized "time" picture of subsurface reflectors. Depth migration uses a detailed interval velocity model in an attempt to reconstruct a distortion-free depth picture. Lateral velocity gradients can cause time migration to distort structures, often beyond the point of correction by depth conversion. When this happens, depth migration after seismic modeling becomes the preferred alternative.

With depth migration, the interval velocity model is used to migrate seismic data directly into the depth domain; post-migration depth conversion is not required. Ray bending is accounted for in the migration process, increasing the demand for an accurate depth velocity model.

Like the previous example, the unmigrated seismic line shown in Figure 16 was acquired across a thrust belt in Wyoming. The picture is quite complex in time, with several structures visible along the line.

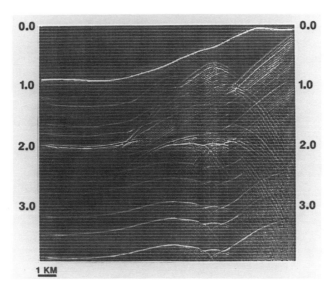

Fig. 14. Synthetic seismic traces generated from the depth velocity model in Figure 13, using *f-x* finite-difference forward modeling. The synthetic seismic traces are a good match with the real data in Figure 11.

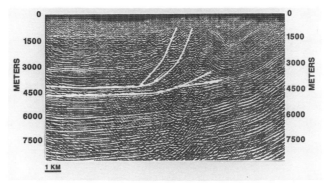

Fig. 15. The final step in the process is depth conversion. The model has helped the processor to produce a more reliable depth image.

Figure 17 shows the synthetic time plot obtained from modeling, overlain on the real data. The yellow-to-blue section has been thrust over the younger gray-to-yellow section. On the time section, four major structures are indicated by arrows. However, as will be shown only *two* of these structures are legitimate. The other two are false structures, created by rapid lateral velocity changes.

In Figure 17, the arrow at the far right points to an anticline between 0.8 and 2.7 s. In actuality, this anticline is artificial, and is caused by a time sag just to the right of the apparent structure. The time sag is caused by irregularities in the near-surface, low velocity material.

Figure 18, the depth velocity model, shows a thickening in the overlying low-velocity material, shown in dark brown at a depth of 1200 m. The low-velocity pod is enough to disturb time reflectors up and down the section. In depth, the anticline isn't there.

Similarly, the deepest arrow in Figure 17, just left of center, points to a large subthrust time structure from 2.2 to 4.0 s. Figure 18, the depth model, shows this to be a subthrust pullup. The false time structure is due to the emplacement of high-velocity material, shown in blue on the depth model.

Using the depth velocity model in Figure 18, the traces can now be migrated directly to depth. Figures 19 and 20 show the depth migration, with and without the velocity boundaries. The two legitimate structures remain, while the two false structures have been corrected.

At this point, a good question is how depth migration can become a more routine process. The key to making depth migration routinely successful is cooperation between the interpreter and processor. Efficient methods for creating

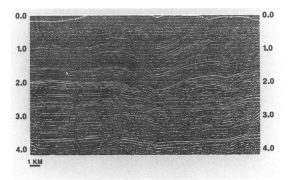

Fig. 16. An unmigrated seismic line acquired over a Wyoming thrust fault. Several potential anticlinal structures can be observed.

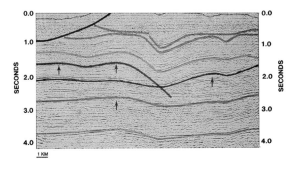

Fig. 17. The synthetic time plot produced from modeling the depth model in Figure 18. Four major "structures" are highlighted with arrows, but the two which are false were created by rapid lateral velocity changes.

and jointly evaluating depth velocity models need to be available to the processing team. When interval velocity models satisfy both geological and geophysical constraints, successful depth migrations can be obtained.

SUMMARY

In conclusion, we have seen a spectrum of lateral velocity contrasts and the problems they pose, ranging from mild to complex. (We have not seen the complex end of that spectrum by any means.) Depth conversion and depth migration are both valuable tools for prospecting in areas with lateral velocity gradients. These techniques can downgrade prospects that should not be drilled, and reveal, or at least better delineate, prospects that should be drilled.

The role of seismic modeling is becoming more important for processing,

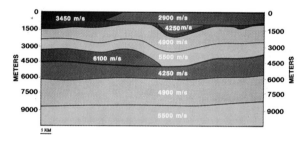

Fig. 18. The final depth model, showing that only two of the four time structures cited in Figure 17 are true depth structures. The false structures in Figure 17 are: the one at far right, and the subthrust structure, left of center. See text for details.

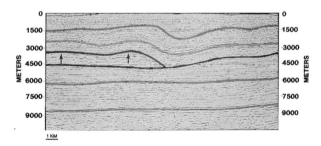

Fig. 19. The depth migration of the traces in Figure 16. This migration was obtained using the velocity distribution in Figure 18. Some velocity boundaries have been overlain for reference.

Fig. 20. The depth migration from Figure 19, shown without overlay. The two legitimate structures remain, both upthrown and left of center, while the two false structures have been corrected.

interpreting, and exploring complex structures. The key to migrating in these areas is an accurate depth velocity model. This model usually requires the involvement of an interpreter and some form of modeling in the processing sequence.

If lateral velocity variations are not too severe, time migration and depth conversion may be an acceptable alternative to depth migration. Ray trace modeling should be used to verify interval velocity models. When structure is complex in three dimensions, 3-D modeling, 3-D seismic data, and map migration are additional, valuable tools to consider.

ACKNOWLEDGMENTS

M. S. "Mitch" Yancey was a leader in promoting and applying the modeling and depth migration concepts demonstrated here. Bob Stolt and Al Benson provided the theoretical tools to make these concepts work. We are grateful to Conoco Inc. for the opportunity to work on this team, and to publish our results.

Thanks also to Stuart Fagin for improving the manuscript, and to Chris Littlecook, Mindy Rich-Herbert, and Joe Hoover for drafting the illustrations.

REFERENCES FOR GENERAL READING

D'Onfro, P. S., Weinberg, D. M., Johnson, J. H., and Yancey, M. S., 1983, Drape fold, South Elk Basin, Wyoming: in Bally, A. W., Ed., Seismic expression of structural styles: Am. Assn. Petr. Geol., 3, 2.2.12–14.
McClellan, B. D., and Storrusten, J. A., 1983, Utah-Wyoming Overthrust line, in Bally, A. W., Ed., Seismic expression of structural styles: Am. Assn. Petr. Geol., 3, 4.1.39–44.
Stolt, R. H., and Benson, A. K., 1986, Seismic migration—theory and practice: in Helbig, K., and Treitel, S., Eds., Seismic exploration series, 5, Geophysical Press.

CASE HISTORY 3

Seismic Modeling of an Imbricate Thrust Structure from the Foothills of the Canadian Rocky Mountains

*Steven Lingrey**

ABSTRACT

Seismic structural modeling provides a useful means toward understanding the complex geometry associated with detached-style fold and thrust deformation. Routine seismic methods do not tightly constrain details of structural interpretation, but methods incorporating geometric and seismic modeling allow these details to be inferred. Geometric models measure spatial elements of the folds and faults, and check them for internal consistency, usually against an assumed condition of material balance. Seismic models measure the effects of ray-path bending through a complex, structurally defined velocity field. The seismic model predicts the presence (or absence) of reflections and their pattern on unmigrated sections. These patterns in the modeled data can be checked against the patterns observed in the real data. Iteration between the two models allows the interpretation to converge toward a mutually acceptable solution.

Structural analysis of a seismic profile across the Quirk Creek gas field from the Foothills belt of the Canadian Rocky Mountains is used to illustrate this iterative method of seismic interpretation. The internal geometry of thrust sheets, essentially opaque on the basis of routine examination of a migrated seismic profile, is developed with the aid of geometric and seismic modeling techniques. As with many model-based approaches, proposed solutions in a given geometric or seismic model are nonunique. Used in combination, however, they each check solutions independently and thus can more narrowly constrain the range of acceptable interpretations. The incorporation of synthetic seismic modeling greatly improves the accuracy of interpretation for the structurally complicated Quirk Creek gas field.

INTRODUCTION

Structural interpretations of reflection seismic data in the Foothills belt of the Canadian Rocky Mountains of southern Alberta have typically relied heavily on concepts of structural style (Bally et al., 1966; Dahlstrom, 1970; Harding and Lowell, 1979; Sheriff and Geldart, 1983). This reliance on structural style is necessary in view of the fact that seismic images of fold and thrust structures are typically incomplete and of uneven quality. Parts of the stratal geometry may be

*Esso Resources Canada Ltd.

clearly shown while other parts show either a lack, or a confusing overabundance of reflection signals. Migrated results of uniform quality are difficult to obtain. Thus, generalizations of structural style have been necessary to complete interpretations in areas where imaging is poor.

In general, characteristic patterns of seismic reflections reveal the large scale (regional) geometry of detached-style fold and thrust deformation (e.g., the several seismic illustrations in Bally, 1983). In detail, however, the internal geometry of individual fold and fault structures is not as well defined. Barring difficulties in acquisition or statics, the regions of poor seismic image can be attributed to the effects of ray-path bending caused by the structure itself. Routine processing typically does not adequately account for this effect. The main purpose of this paper is to show how synthetic seismic modeling provides a means for assessing this effect and improving the interpretation.

A structural interpretation of a seismic profile across the Quirk Creek gas field from the Foothills belt of the southern Canadian Rocky Mountains is presented (Figure 1). High effort seismic processing (refraction statics corrections, velocity sweeping for optimal VNMO, and recursive migration) provides a high quality data base. Yet characteristically, the seismic reflections incompletely image the structure due to the complexity of folds and faults. Standard interpretation methods are augmented with the inclusion of geometric and seismic modeling analysis. Geometric models are incorporated to test the interpretation for material balance. Seismic models are incorporated to test the interpretation for reflection imaging potential.

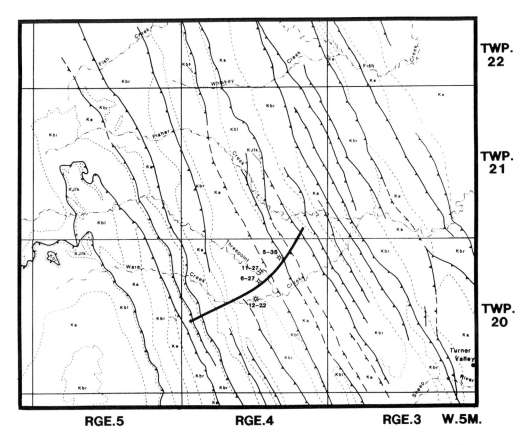

Fig. 1. Location map showing the surface geology for the Foothills Belt of the Canadian Rocky Mountains southwest of Calgary, Alberta and the location of the seismic profile.

METHOD OF SEISMIC INTERPRETATION

Since full seismic imaging of the complicated structural geometry associated with fold and thrust deformation in the Quirk Creek structure cannot be produced, routine seismic interpretation methods are insufficient to achieve an accurate interpretation. Two modeling procedures provide an opportunity to remedy this limitation (Figure 2). First, a geometric model of the seismic interpretation can be constructed and this depth image can be analyzed for material balance. Second, a seismic model (using the balance-tested geometric model) can be run to predict the ray-path distortion and its effect on the patterns of reflections. Modifications in the structural geometry of the interpretation can be tried and assessed on the basis of the fit between observed and predicted (i.e., modeled) reflection patterns. Iteration between the geometric and seismic modeling allows modifications to converge on the strongest interpretation. In the end, the structural interpretation is more completely constrained by the seismic data, even in areas where reflections are missing or improperly migrated.

Geometric modeling involves measurements of certain spatial elements, generally to check for a condition of material balance. In its simplest form, a balanced section assumes conservation of bedlength and this status can be tested by measuring, and comparing for equality, successive pretectonic stratal horizons. In a more advanced form, a kinematic analysis is used to compare pre- and post-deformational profiles. A forward model starts with an undeformed profile and computes the deformed state; an inverse model (or palinspastic restoration) starts with a deformed profile and computes the undeformed state. For a two-dimensional (2-D) analysis, plane strain is assumed. In this study, a palinspastic restoration is used to demonstrate the material balance of the structural interpretation.

The mechanism of deformation is important to geometric modeling as this defines the mathematical process by which the strain is calculated. In this

PROCEDURE FOR INTERPRETATION

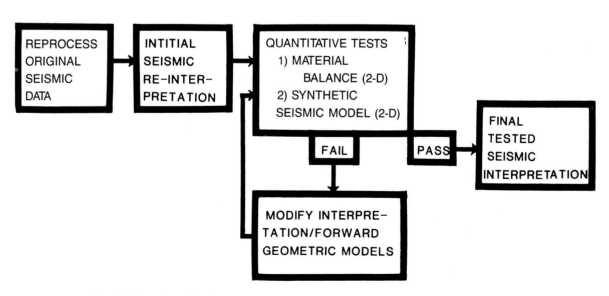

Fig. 2. Flow chart for the method of interpretation in areas of complex structure.

analysis, a flexural-slip mechanism is used to inversely model the folding deformation (Donath and Parker, 1964). As such, bedding surfaces approximate planes (lines in 2-D) of no finite strain within the strain ellipsoid. The principal constraints on flexural-slip deformation on the geometry of folds in fold and thrust belt settings were outlined in Dahlstrom (1969). Recently, Suppe (1983; 1985) has trigonometrically defined two mechanisms of detached-style, flexural-slip folding and thrust faulting: fault-bend and fault-propagation folding. In the process of inferring structural interpretations, the patterns of folds and faults were calculated by means of graphic experiments using Suppe's formulations (forward models).

Geometric models were constructed and restored with the aid of commercially available computer software (Kligfield et al., 1986). This software enables precise representation and kinematic restoration of flexural-slip geometry.

Synthetic seismic modeling involves the analysis of ray-path bending through a complex velocity field (Taner et al., 1970; May and Hron, 1978). Because the pattern of reflections in a time display is sought, and not their amplitude character, the model profile defines relatively thick layers representing averaged interval velocities. The object of the analysis is to calculate the traveltime of a zero-offset ray-path to and from a given reflective horizon. This can be plotted as a synthetic time section showing the pattern of reflection arrivals for an unmigrated, stacked section.

The synthetic result can be compared with the reflections present in the real section. Two modifications to improve the fit between synthetic and real data are possible: (1) the geometry of the layer boundaries in the model can be changed, or (2) the average interval velocities can be changed. In this study, there are several wells containing velocity information on the Mesozoic and Paleozoic rock units; the average sonic velocities used are displayed in Table 1. The values of these velocities are relatively uniform and not too dependent on depth. This uniformity allows the modification of the seismic model to critically test, principally, for the geometry of the layer boundaries.

Seismic models were constructed using the tested geometric models. These depth profiles were input into an interactive modeling system for computation of ray-tracing and plotting of resultant reflection arrivals on synthetic time sections.

STRUCTURAL INTERPRETATION

Figure 3 is a migrated seismic profile situated across the southern part of the Quirk Creek gas field located near the external margin of the Foothills belt of the Canadian Rocky Mountains of Alberta. The approximate location of the line is shown in Figure 1. The structure at Quirk Creek is a thrust faulted, anticlinal trap involving imbricate repetitions of upper Paleozoic carbonate rocks. This Paleozoic level structure is not represented at the surface. No upper Paleozoic rocks crop out at the surface and no conspicuous pattern of fold closure appears in overlying Cretaceous strata at the surface. A stratigraphic chart shows the formations, lithology, and average interval velocities for the sedimentary rocks in this part of the Foothills (Figure 4).

Table 1. Interval velocities used for depth conversion and seismic modeling

Stratal units	Interval velocity
Upper Cretaceous/Lower Cretaceous	13,600 kft/s (4.15 km/s)
Lowermost Cretaceous/Jurassic	15,000 kft/s (4.57 km/s)
Mississippian	21,500 kft/s (6.55 km/s)
Devonian/Cambrian	21,000 kft/s (6.40 km/s)

The goal of this seismic interpretation aims at a more detailed description of the Paleozoic structural geometry of the Quirk Creek trap. The analysis utilizes: (1) the recently recognized principals of structural deformation in detached fold and thrust terranes (Dahlstrom, 1970; Boyer and Elliot, 1982; Suppe, 1983, 1985; Woodward et al., 1985), particularly the importance of the presence of several bedding-parallel detachments (duplex style imbrication), (2) an evaluation of material balance by means of computer-aided geometric modeling (Hossack, 1979; Ramsay and Huber, 1987; Kligfield et al., 1987; Marshack and Woodward, 1988), and (3) an evaluation of the final depth interpretation by means of seismic structural modeling.

Seismic interpretation of the line across the southern part of the Quirk Creek structure was aided by four wells and surface geology taken from Geological Survey of Canada maps (Figure 1; Hume, 1931, 1941; Hume and Beach, 1942; Hage, 1946). The results of the structural interpretation are illustrated in Figure 5.

Structural interpretation of the seismic data is influenced by the accepted importance of bedding-parallel detachment faults. For this part of the Foothills and at the structural level of the Quirk Creek structure, four stratigraphic horizons of detachment are important (Figure 5): (1) a basal detachment near the base of the Cambrian, (2) a detachment within the lower part of the Mississippian Banff Formation, (3) a detachment within the Jura-Cretaceous Fernie-Kootenay Formations, and (4) a detachment near the top of the Lower Cretaceous Blairmore Group. Additional detachments within the Upper Cretaceous section are important, but not critical to the interpretation presented.

The importance of multiple surfaces of bedding detachment is shown in Figure

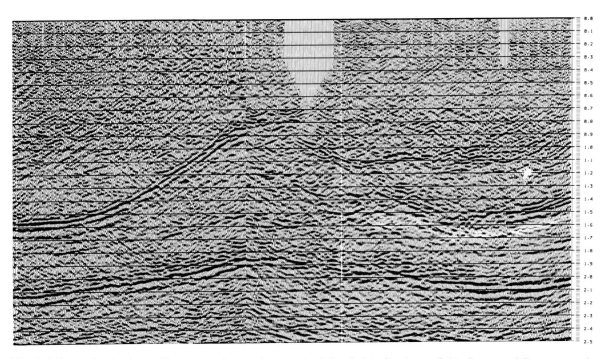

Fig. 3. Migrated seismic profile across the southern end of the Quirk Creek gas field. Structural features at the Paleozoic level are outlined by: (1) a band of high amplitude reflections (Jura-Cretaceous, near top of Mississippian) located at 1.6 s on the left edge, continuous and rising to 0.8 s in the center, and discontinuous and returning to 1.5–1.6 s on the right edge, and (2) a band of high amplitude, lower frequency reflections (mid-Cambrian) continuous across the length of the section located at 2.2 s on the left edge, gently rising to 1.9 s in the center, and located at 2.1 s on the right edge. The upper Paleozoic is imbricately faulted by thrusts, but the lower Paleozoic is autochthonous and unfaulted. Note also the band of high amplitude reflections between 1.0 and 1.1 s on the right half of the seismic line. These reflections arise from the same stratal reflectors described in (1), but at this location do not overlie the Mississippian. The base of these reflections marks the Fisher Mountain thrust fault which overlies repeated Lower Cretaceous Blairmore Group strata.

AGE		FORMATION OR GROUP	LITHOLOGY	THICKNESS	AV. VELOCITIES IN ft/s
CRETACEOUS	UPPER	EDMONTON	UPPER BRAZEAU	1500 - 2000	11,000 / 12,000
		BEARPAW		0 -150	
		BELLY RIVER	LOWER BRAZEAU	1200 - 3500	12,000 / 13,000
		WAPIABI	ALBERTA GROUP	1250 - 1500	12,500 / 13,000
		CARDIUM		300 - 400	13,000- 14,000
		BLACKSTONE		800 - 1000	13,000 / 14,000
	LOWER	BLAIRMORE		1100 - 1400	14,000 / 16,000
		KOOTENAY		0 - 1200	14,000-16,000
JURASSIC		FERNIE		50- 500	13,500-14,000
MISSISSIPPIAN		MOUNT HEAD	RUNDLE	200 - 300	
		TURNER VALLEY		250 - 350	
		SHUNDA		250 - 300	21,000
		PEKISKO		250 - 300	
		BANFF		500 - 600	
DEVONIAN	UPPER	PALLISER		800 - 1000	21,000
		FAIRHOLME		1000 - 1300	
CAMBRIAN	?U?	? ARCTOMYS ?		0 - 100?	
		PIKA		150 - 400	
	MIDDLE	ELDON		500 - 900	20,000
		STEPHEN		200 - 250	
		CATHEDRAL		200 - 800	
PE		HUDSONIAN			

Fig. 4. Table of formations for the southern Alberta Foothills of the Canadian Rocky Mountains showing sites of bedding detachment and average interval velocities. Paleozoic interval velocities are insensitive to depth of burial; Mesozoic interval velocities only mildly increase with depth of burial. (modified from Gordy et al., 1975).

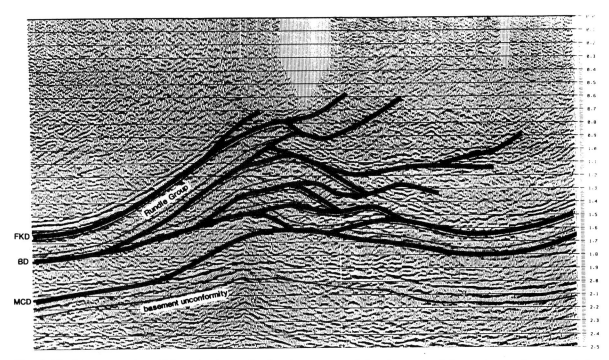

Fig. 5. Final interpretation of the seismic profile showing structural development of thrust sheets above Cambrian and Banff detachments. Structural deformation above the Fernie-Kootenay and Blairmore detachments is not interpreted. Time converted well control is from wells listed in Appendix 1.

6. Many interpretations have emphasized a basal detachment, but have neglected or minimized the higher detachments. The structural interpretation presented here for the Quirk Creek structure emphasizes a duplex-style imbrication of the Mississippian strata between the Banff and Fernie-Kootenay detachments.

The gross features of the trap image well on routinely processed seismic data as a band of three to four, high amplitude reflections that mark a position near the base of the Blairmore Group and the top of the Kootenay Formation. While not precisely at the top of the Paleozoic (Mississippian Mount Head Formation), these reflections are close enough (25–50 ms) to reliably delimit the top of Paleozoic at most places. These reflections define an asymmetric, northeast-vergent fold in profile. The seismic dip line (Figure 3) shows the top of Paleozoic as deep on the southwest (1.6–1.7 s), high on the crest (0.8–0.9 s), then deep again on the northeast (1.55–1.65 s). The structure shows a smoothly defined crest and southwest-dipping backlimb, and a raggedly defined, northeast-dipping forelimb. Reflections do not explicitly define the shape of the forelimb, yet do constrain the region of its occurrence. While much of the external aspect of the structural culmination is clearly discernible, several deep wells into the structure demonstrate a complicated internal structure consisting of several repetitions of the Mississippian section. This internal structure is *not* clearly imaged in the seismic profile and thus cannot be interpreted unambiguously.

A deeper, single reflection of high amplitude marks a position near the base of the Banff (detachment). The continuity of this reflection is poor, but the time dip clearly shows the reflection to be more flat-lying than the top of Mississippian. This reflection pattern is indicative of a floor thrust (lower detachment) above which the overlying strata have been tilted by imbrication. This reflection shows a long wavelength anticlinal warp on the time section, but with a relatively small structural relief (220–260 ms). A small wavelength step in this reflection occurs just east of the crest of the gentle anticline. As is discussed later, the gentle anticline is an artifact (velocity anomaly) on the time section. The sharper,

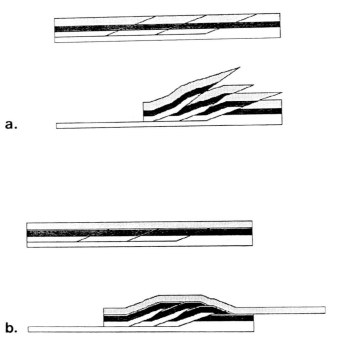

a.

b.

Fig. 6. Structural style of imbrication involving: (a) single basal bedding detachment, and (b) two levels of bedding detachment (duplex style imbrication). Note the absence of tectonic thickening on the left and right edges of the diagram.

east-descending step is interpreted to be the fold at the leading-edge of a thrust coming from the deeper, basal detachment.

An even deeper band of two to three, high amplitude (lower frequency) reflections mark a position near the upper part of the Cambrian section. In contrast to the near top Paleozoic reflections and the near base of Banff reflection, these Cambrian reflections are more or less continuous, indicating little or no disruption by thrust faults (i.e., these strata are largely below the basal detachment and the detachment is interpreted to lie at the top of this package of reflections). These reflections also display a gentle anticlinal warp on the time section with a structural relief of 200–250 ms.

The pattern of these prominent reflection bands demonstrates unambiguously that the Quirk Creek structure is a product of duplex style imbrication. The Fernie-Kootenay detachment (the base of the upper reflections band) is the folded roof thrust of the duplex. The duplex is relatively simple being confined for the most part to imbrications between the Banff detachment (floor thrust) and the Fernie-Kootenay detachment (roof thrust). In detail, the deeper duplex imbrication between the Cambrian and Banff detachments slightly modifies (folds) the shallower duplex.

While the recognition and tracing of the external shape of the duplex deformations can be made with some confidence, the internal geometry of the individual thrust sheets comprising the duplex are less clearly defined. Two types of reflection patterns are critical for the definition of these thrust sheets (Figure 7). The first type of pattern is that associated with the reflection terminations marking the leading edge of a thrust sheet. These terminations indicate the position of the hanging-wall ramp. The second type of pattern is that associated with the reflection terminations marking the trailing edge of a thrust sheet. These terminations indicate the position of the foot-wall ramp. Many reflection patterns on the dip lines are suggestive of these structural elements. To confirm or deny these inferred thrust ramps and the overall structural interpretation, two tests have been made in the process of achieving the final interpretation. These tests are: (1) geometric modeling that involved checks on 2-D material balance (palinspastic restoration), and (2) seismic structural modeling that involved zero-offset ray tracing to confirm the magnitude of pull-up effects and to predict imaging attributes on unmigrated seismic data.

GEOMETRIC MODEL

Initial structural interpretations of the seismic data are tested by means of geometric models. This process involves: (1) depth conversion of the seismic interpretation, (2) geometric analysis of the interpretation (checks for material

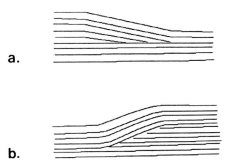

Fig. 7. Line drawing showing the pattern of reflections that are indicative of: (a) leading-edge of thrust sheet, and (b) trailing-edge of thrust sheet.

balance), and (3) modifications of the interpretation to remedy areas of imbalance in the interpretation. The computer program allowed the impact of changes in the geometry of the interpreted thrust sheets to be assessed, thereby suggesting changes that could improve areas of imbalance. For the Quirk Creek seismic line, three cycles of interpretation modification were involved.

Depth conversion of the interpreted seismic line in this part of the Foothills Belt of the Canadian Rocky Mountains is aided by: (1) the platformal character of the Paleozoic strata which shows a very gradual primary thickening (approximately 0.3 degrees) of stratigraphic layers to the west, (2) the fairly uniform interval velocities characteristic of the Mesozoic strata, and (3) deep well control, one with check-shot information.

To the northeast and southwest of the Quirk Creek structure, there are zones in which the Paleozoic has not been duplicated across thrust ramps. In these zones, the top of the Paleozoic in depth can be calculated using the average interval velocity for Mesozoic strata treated as one layer. This is fortunate, because the necessity to recognize and compensate for the complexly deformed Mesozoic layers is obviated. By connecting the calculated depth in front (northeast) of the Quirk Creek structure to the calculated depth in back (southwest), a regional gradient in depth for the top of the Paleozoic is established. Using known thicknesses of Paleozoic strata (checked for internal consistency with the time thickness in the seismic interpretation), the regional gradient of the autochthonous basement can be determined.

Estimations of the appropriate depths across and within the Quirk Creek structure can be either calculated downward using the interval velocities, or calculated upward using the net amount of duplicated Paleozoic section above the regional gradient. These calculations are well constrained; since for each individual thrust sheet interpreted, the horizontal dimension of its leading- and trailing-edge can be measured directly and the vertical dimension between the Banff and Fernie-Kootenay detachments can be inferred from the nearly uniform stratigraphic thickness. Forward geometric models using the idealized fault-bend fold behavior were used in some cases to anticipate the spatial details of imbricate thrust patterns (Suppe, 1983).

The depth conversion of the initial interpretation made for the Quirk Creek seismic line is shown in Figure 8 with its palinspastic restoration. Significant gaps occur between the restored thrust sheets indicating a material imbalance in this interpretation. Accepting the appropriateness of the assumptions (plane strain, flexural-slip folding), the implication of this analysis is that the interpretations need to be modified to remove the spatial gaps between the thrust sheets. As the external geometry of the Quirk Creek structure is rather tightly constrained by the seismic data, modifications are directed toward the internal geometry of the thrust sheets.

In the following discussion the seismic model for this geometric model points out a significant discrepancy between the calculated and observed time anomaly in unfaulted Cambrian reflections underlying the structure. The implication of this analysis is that too much Paleozoic and not enough Mesozoic is interpreted in the core of the structure.

The depth conversion of the final interpretation made for the Quirk Creek seismic line is shown in Figure 9 with its palinspastic restoration. This result follows two reinterpretations in which modifications were made and tested to see if the state of material balance improved. The restoration shows no significant gaps. While not necessarily a unique solution, this interpretation is demonstrably consistent with the assumed mechanism of deformation. When joined with the observation that the seismic model for this interpretation well matches the observed seismic reflection patterns in the unmigrated section, the accuracy of the final interpretation is considered to be optimized.

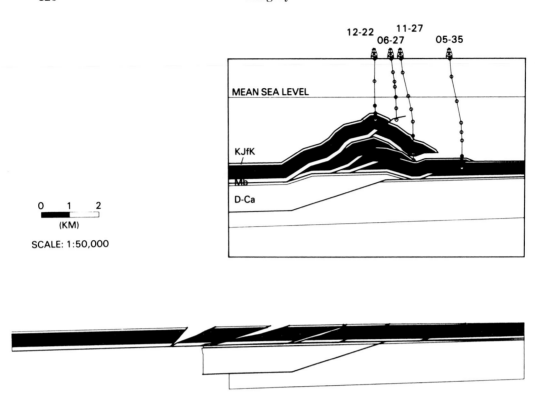

Fig. 8. Depth converted structural profile of initial interpretation for seismic line across southern Quirk Creek gas field showing palinspastic restoration.

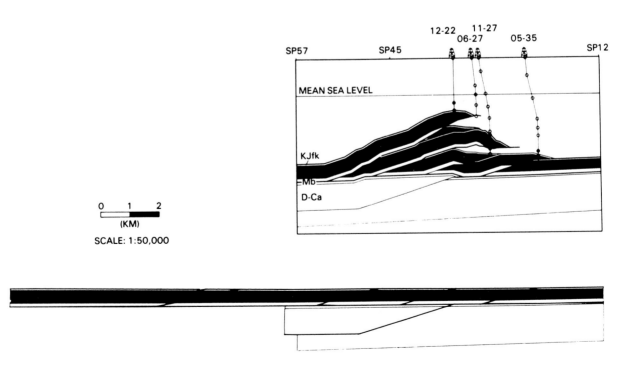

Fig. 9. Depth converted structural profile of final interpretation for seismic line across southern Quirk Creek gas field showing palinspastic restoration.

SEISMIC MODEL

Initial structural interpretations of the seismic data, in addition to the geometric models, are tested by means of synthetic seismic models. This process involves: (1) depth conversion of the seismic interpretation (the same depth converted model used for geometric analysis can be input for synthetic seismic analysis), (2) definition of model layers and their appropriate average interval velocities, (3) computation of the zero-offset ray-paths using the general acquisition parameters (seismic datum, shot-point spacing, etc.) of the real seismic line, (4) generation of a plot of reflection arrivals at the proper time scale for comparison with the real seismic line, and (5) modification of the interpretation to remedy areas of mismatch in the unmigrated reflection image. Small modifications to one or two of the layer geometries in the model allow the impact of interpretation changes to be assessed. These modifications could be tried interactively using the modeling program. For the Quirk Creek seismic line, two of the three geometric models generated were input as seismic models.

The seismic model of the initial interpretation made for the Quirk Creek seismic line is shown in Figure 10 showing the average interval velocities assigned the 18 layers. The model calculates reflection arrivals for the top of Paleozoic, the top of the Devonian (equivalent to the Banff detachment), and the base of Cambrian (about 150 ms below the Cambrian reflections). The predicted reflection patterns are shown in Figure 11, overlaying the unmigrated seismic data. As is the case with the geometric model of this interpretation, several problems are apparent. An important attribute of the model is the strong time "pull-up" of the spatially flat-lying top of basement reflector. This attribute constitutes the most conspicuous discrepency between the predicted and observed reflections. The model predicts a maximum time pull-up of 300 ms, while the observed pull-up of Cambrian reflections is 220 ms.

Estimation of the expected pull-up effect due to the imbrication of high velocity Paleozoic carbonates (interval velocity = 6.4–6.5 km/s; 21.5 kft/s) into lower velocity Mesozoic clastics (interval velocity = 4.11–4.27 km/s; 14.0 kft/s) is shown in Figure 12. Imbrication result from displacement along thrust faults repeats part or all of the upper Paleozoic section. The repetition creates a tectonically thickened section in the region of the thrust ramp; up-dip and down-dip of the ramp, the section may show normal stratigraphic thickness. As such, traveltimes beneath a region of thickened carbonate strata are less than those beneath a region

INITIAL QUIRK CREEK MODEL FOR LINE 60636

Fig. 10. Seismic model of initial interpretation (same structural interpretation shown in Figure 8) showing average interval velocities of model layers.

of normal thickness for an equivalent depth. Thus a repetition of 550 m (1800 ft) of Paleozoic carbonate create the effect of lowering the traveltimes (velocity pull-up) for reflections beneath the repetition by 90–100 ms (Figure 12). Duplications of Lower Cretaceous clastics (interval velocity of some layers as high as 4.57 km/s; 15,000 kft/s) can also cause velocity pull-up, but the magnitude of repetition must be over three times that of a Paleozoic repetition to produce an equivalent effect. Modifications to the initial interpretation that include more Mesozoic strata

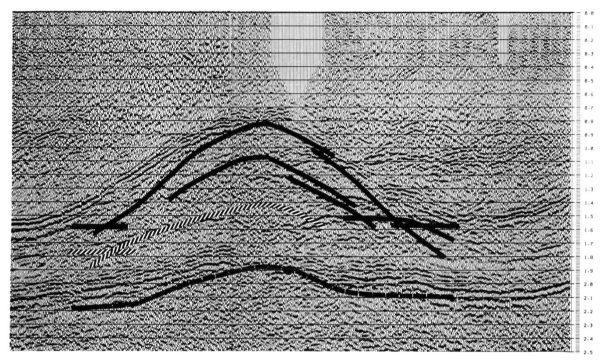

Fig. 11. Plot of reflection arrivals calculated for zero-offset ray-paths through model shown in Figure 10.

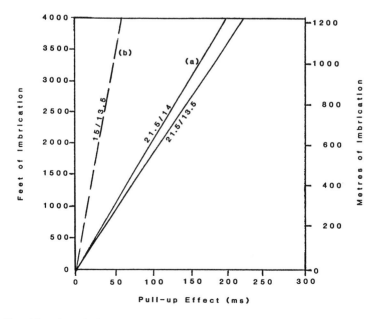

Fig. 12. Graphic plot of the calculated magnitude of time pull-up when replacing: (a) high-velocity carbonates and (b) high-velocity clastics.

in internal imbrication of the Quirk Creek structure should improve the fit of the model.

In the process of seismic reinterpretation, the graph shown in Figure 12 was used to estimate the amount of pull-up for any interpreted amount of Paleozoic imbrication. Excess Paleozoic thickness (the observed tectonic thickening in time of Paleozoic strata) was measured and the calculated pull-up was determined. The consistency between the structural interpretation and the observed velocity pull-up effect was thereby maintained. Changes in reinterpretation needed to preserve the magnitude and location of the pull-up.

The seismic model for the final interpretation is shown in Figure 13. This model incorporates reinterpretations in which modifications were checked against material balance and estimations of velocity pull-up. The predicted pattern of reflection arrivals now constitutes a good match with the real data (Figure 14). In addition to the velocity pull-up simulated in the Cambrian and Devonian reflections, the patterns characterizing the near top Mississippian reflections approximate the seismic image observed. To reiterate, the accuracy of this final interpretation is considered optimal, given the harmony between results of the independent checks of the geometric and seismic models.

To contemplate more sophisticated seismic models, and to recognize that tighter constraints on the interpretation are possible, is interesting. Gather modeling could check the quality of VNMO correction and the adequacy of assumed hyperbolic moveout patterns. Diffraction arrivals could be plotted to identify diffraction signals in real data that are not adequately suppressed in stacking and migration. Migration of synthetic sections could evaluate the effectiveness of the migration algorithm on the data set. Finally, 3-D seismic models could predict sensitivity to 2-D aquisition methods.

CONCLUSIONS

Interpretations of fold and thrust belt structures on reflection seismic data have frequently broken down when tested by well penetrations. The routine methods of aquisition, processing, and interpretation seem to be insufficient to the problem of predicting the often complex structural geometry. Geometric and seismic models provide a means for extending routine methods to cope with the complex geology.

A geometric model can bring in constraints on an interpretation in the form of strain analysis. Certain geometric attributes of the interpretation, appropriate for

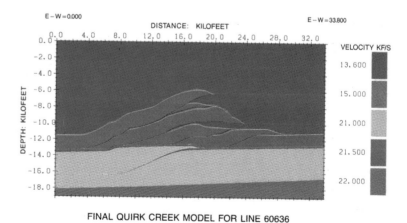

FINAL QUIRK CREEK MODEL FOR LINE 60636

Fig. 13. Seismic model of final interpretation (same structural interpretation shown in Figure 9) showing average interval velocities of model layers.

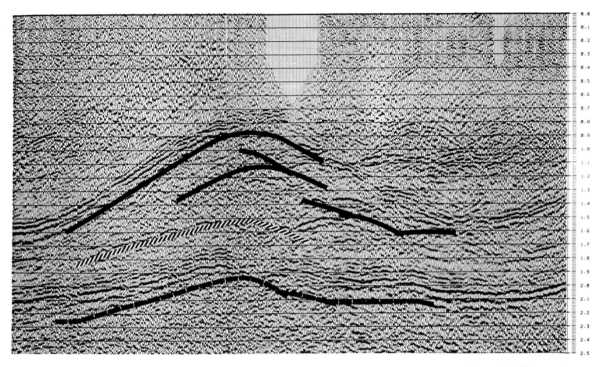

Fig. 14. Plot of reflection arrivals calculated for zero-offset ray-paths through model shown in Figure 12.

the assumed mechanism of deformation, can be checked and if need be, modified. A seismic model can bring in constraints on an interpretation in the form of ray-path analysis. The imaging potential of unmigrated, stacked sections can be predicted. Provided the interval velocity properties of the sedimentary rocks being deformed are known, the seismic model predicts the effect of interpreted structural geometry on the ray-paths. Seismic attributes that may or may not be constructive to the standard reflection image can be identified.

Each of the models can lead to solutions that are not necessarily unique. Each model, however, works independently. Used in combination the models can more narrowly constrain the range of acceptable solutions. The opportunity therefore exists to greatly improve the accuracy of interpretation in structurally complicated areas.

REFERENCES

Bally, A. W., 1983, Seismic expression of structural styles, Volume 3—Tectonics of compressional provinces/strike slip tectonics: Am. Assn. Petr. Geol. Studies in Geology Series No. 15.

Bally, A. W., Gordy, P. L., and Stewart, G. A., 1966, Structure, seismic data and orogenic evolution of the southern Canadian Rocky Mountains: Bull. Can. Petr. Geol., 14, 337–381.

Boyer, S. E., and Elliott, D., 1982, Thrust systems: Bull., Am Assn. Petr. Geol., 66, 1196–1230.

Dahlstrom, C. D. A., 1969, Balanced cross sections: Can. J. Earth Sci., 6, 743–757.

———, 1970, Structural geology in the eastern margin of the Canadian Rocky Mountains: Bull. Can. Petr. Geol., 18, 332–406.

Donath, F. A. and Parker, R. B., 1964, Folds and folding: Geol. Soc. Am. Bull., 75, 45–62.

Hage, C. O., 1946, Dyson Creek: Geol. Surv. Canada Map 827A.

Harding, T. P. and Lowell, J. D., 1979, Structural styles, their plate tectonic habitats, and hydrocarbon traps in petroleum provinces: Bull., Am. Assn. Petr. Geol., 63, 1016–1058.

Hossack, J. R., 1979, The use of balanced cross-sections in the calculation of orogenic contraction: A review: J. Geol. Soc. London, 136, 705–711.

Hume, G. S., 1931, Turner Valley sheet: Geol. Surv. Canada Map 257A.

————, 1941, Fish Creek: Geol. Surv. Canada Map 667A.

Hume, G. S. and Beach, H. H., 1942, Bragg Creek: Geol. Surv. Canada Map 654A.

Kligfield, R., Geiser, P., and Geiser, J., 1986, Construction of geologic cross sections using microcomputer systems: Geobyte, Spring issue, 60–66.

Marshack, S. and Woodward, N., 1988, Introduction to cross-section balancing: in Marshack, S. and Mitra, G. (eds.), Basic methods of structural geology: Prentice-Hall, Inc., 303–332.

May, B. T. and Hron, F., 1978, Synthetic seismic sections of typical petroleum traps: Geophysics, 43, 1119–1147.

Ramsay, J. G., and Huber, M. I., 1987, The techniques of modern structural geology, Volume 2, Folds and fractures: Academic Press, 309–700.

Sheriff, R. E. and Geldart, L. P., 1983, Exploration seismology, Volume 2, Data-processing and interpretation: Cambridge University Press, 221.

Suppe, J., 1983, Geometry and kinematics of fault-bend folding: Am. Jour. Sci., **283**, 684–721.

————, J., 1985, Principals of structural geology: Prentice-Hall, 537.

Taner, M. T., Cook, E. E., and Neidell, N. S., 1970, Limitations of the reflection seismic method, Lessons from computer simulations: Geophysics, **35**, 551–573.

Woodward, N., Boyer, S. E., and Suppe, J., 1985, An outline of balanced cross-sections: Univ. Tenn. Dept. Geol. Sciences Studies in Geol. 11, 2nd ed., 170.

CASE HISTORY 4

Seismic Modeling of Fault-Related Folds

Peter F. Morse*, Guy W. Purnell*, and Donald A. Medwedeff[‡]

ABSTRACT

We present examples from a class of geologic models for compressional tectonic regimes along with their corresponding seismic expressions on unmigrated and migrated seismic profiles. The geologic models are based upon fault-related fold theory (Suppe 1983, 1985). For most of our seismic examples, we use modeling and migration programs based on the acoustic wave equation. Such wave-equation techniques, while not as computationally fast, often handle complex models more realistically than geometric raytracing and provide output better suited for subsequent data processing.

Synthetic 2-D zero-offset sections, computed using the wave-equation exploding-reflector approach, lead to (1) recognition of patterns associated with different fault-related folds, and (2) prediction of some of the difficulties in working with unmigrated and migrated seismic sections. Synthetic multioffset shot records, generated using the 2-D acoustic wave equation, demonstrate difficulties in imaging beneath fault-propagation folds using the conventional CMP method. A synthetic 3-D zero-offset survey across a 3-D fault-bend fold demonstrates that conventional 2-D seismic lines acquired in the fault-slip direction are insufficient for correct structural imaging and interpretation; hence, 3-D data acquisition and processing are necessary.

Finally, we examine a case history from Lost Hills field, San Joaquin Valley, California. This study demonstrates that geologically plausible interpretations consistent with both well and seismic data can be generated by iterating between geologic interpretations and synthetic zero-offset sections.

INTRODUCTION

In seismic exploration, subsurface reflectors are commonly imaged using a common-midpoint (CMP) stack, followed by a time-migration of the CMP stack. If sufficient velocity information is available, a depth-migration of the CMP stack is sometimes generated. To arrive at a correct geologic interpretation from such seismic data, an explorationist must have a clear understanding of the relationship between a geologic model and the model's various seismic expressions. We use the term "seismic expression" to denote the seismic response of a geologic model

*Texaco Inc., E&P Technology Division, 3901 Briarpark, Houston, TX 77042.
[‡]ARCO Oil and Gas Company, 2300 West Plano Parkway, Plano, TX 75075. Formerly Department of Geological Sciences, Princeton University, Princeton, NJ 08544.

as the response appears on processed sections such as the CMP stack, the time-migration of the CMP stack, and the depth-migration of the CMP stack. This understanding is especially necessary for interpretation of complex models, which are almost always imperfectly imaged by most of their seismic expressions. For example, the time-migration of a CMP stack, in the presence of significant lateral velocity variations, will not migrate reflections to their proper positions even when the velocities are well known (Judson et al., 1980). Further complicating matters is the limitation that most seismic data are acquired using 2-D field geometries, while the structures being imaged are 3-D.

Using a series of models whose geometry is predicted by fault-related fold theory (Suppe, 1983, 1985), we compute various seismic expressions. (Fault-bend and fault-propagation folds are geologic models proposed to explain some common compressional structures.) The purpose of this computation is two-fold: (1) to illustrate the seismic expressions of fault-bend and fault-propagation folds and (2) to demonstrate the strengths and limitations of certain standard processing techniques (e.g., CMP stack, time migration, and depth migration) in the presence of complex structure and lateral velocity variations. All time and depth migrations discussed in this paper use the finite-difference wave-equation algorithm which requires interval velocities as input.

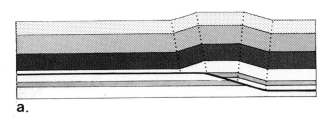

a.

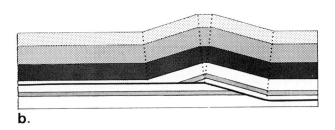

b.

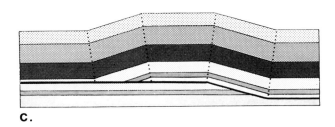

c.

Fig. 1. Three progressive stages in the geologic development of a simple fault-bend fold (Suppe, 1983). As fault-slip increases, thrusting carries the upper sheet across a step in decollement; the layers in that sheet bend at axial planes (dotted lines). (a) Fold amplitude and width growing. (b) End of amplitude growth. (c) Fold width growing.

Instead of ray-trace modeling, in most cases we use finite-difference acoustic wave-equation modeling to create the synthetic seismograms for the geologic models. This choice is based upon the difficulties in tracing rays through the laterally complex models, easy in-house accessibility of wave-equation modeling programs that better handle such models, and resulting output that is better suited for subsequent data processing.

Representative 2-D models are examined using synthetic zero-offset seismic sections and time-migrations of the sections. For one of these models, a synthetic seismic survey consisting of multioffset shot records is computed and processed through CMP stack. Comparison of this CMP stack (before and after migration) to the corresponding zero-offset section (before and after migration) demonstrates some limits to zero-offset modeling and the CMP method. Vertical slices of a 3-D volume of zero-offset data generated for a 3-D overthrust model are compared to vertical slices through the 3-D time-migrated data volume and to corresponding slices of 2-D time-migrated data. This comparison illustrates some pitfalls associated with interpreting data across such a complex structure, even when the seismic lines are oriented in the fault-slip direction. Finally, we present a case

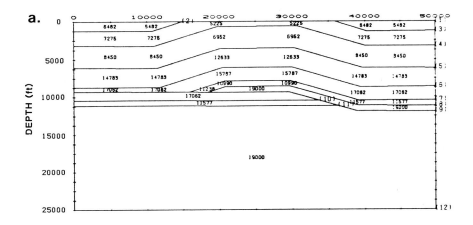

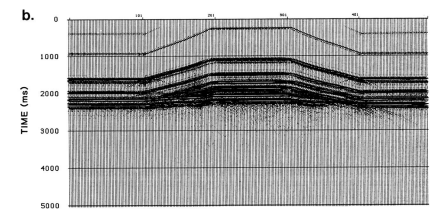

Fig. 2. (a) Velocity model assigned to the final structural model of Figure 1c. Velocity varies with both lithology and depth (compaction). Between the annotated points, velocities are linearly interpolated. (b) Synthetic zero-offset section computed by an exploding-reflector wave-equation algorithm. The effects of lateral velocity variation are not too severe, so interpretation is straightforward.

history which compares the real 2-D CMP stack with synthetic zero-offset sections. The initial geologic model used to generate the first synthetic is modified, according to geologic constraints, to increase the similarity between the real and synthetic sections. This modification leads to a more acceptable interpretation and the dominant features of the modified synthetic closely resemble the stack.

2-D ZERO-OFFSET ACOUSTIC MODELING

Using an exploding-reflector wave-equation method, forward modeling is applied to three 2-D geologic models to create synthetic zero-offset seismic sections, which are then time-migrated. The exploding-reflector concept depicts a seismic experiment in which each reflection point on each reflector in the subsurface explodes at time $t = 0$ with a magnitude proportional to the reflection coefficient. By Huygens' principle, the superposition of all the upgoing waves from the point explosions appears, for a flat interface, like a single wave traveling up from the interface. If the wave propagates at half the actual velocity, the arrival

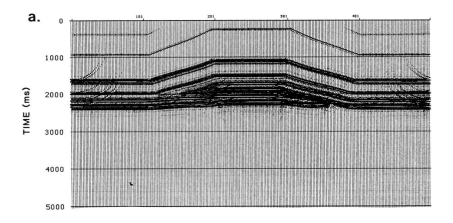

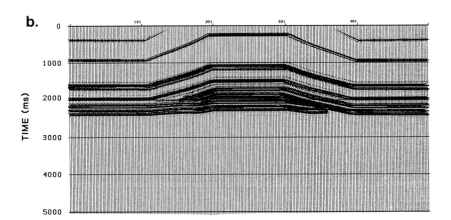

Fig. 3. (a) Time migration of the zero-offset section (Figure 2b) for the fault-bend fold model. Diffractions are now focused, but the structural picture is little changed from Figure 2b. (b) Time model (depth model from Figure 2a converted to two-way vertical time). Note similarity to the time migration, suggesting that time migration is adequate for correctly imaging simple fault-bend fold structures.

times at the recording surface will result in a time section that resembles a zero-offset seismic section. This type of 2-D synthetic section is quick and easy to generate and depicts normal-incidence reflections (those having identical down-going and upgoing wave paths). The exploding-reflector concept does not account for zero-offset data where the reflections are non-normal incidence (c.f. Kjartansson and Rocca, 1979), which may result from lateral velocity variations. Our exploding-reflector algorithm is based on a modification of the 2-D acoustic wave equation that allows waves to propagate upward only. Thus, multiple reflections are not predicted.

Wave-equation synthetics not only appear more realistic than ray-trace synthetics, but are also better suited for further processing as if they were real data. The synthetics in our examples are useful in (1) learning to recognize patterns associated with different types of structures and (2) predicting some of the difficulties to be anticipated when interpreting migrated and unmigrated seismic sections across a given structure.

One way of predicting the effectiveness of time migration is to compare a time-migrated synthetic with the time model corresponding to the input geologic model. (A time model is a geologic model in which the vertical position of a horizon corresponds to the two-way vertical traveltime to the horizon, rather than to the horizon's depth.)

The compressional structures we forward model here are a simple fault-bend

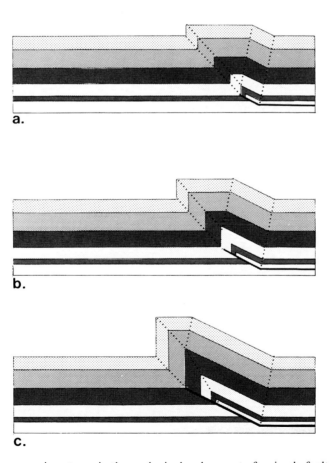

a.

b.

c.

Fig. 4. Three progressive stages in the geologic development of a simple fault-propagation fold (Suppe, 1985). A propagating fault tip angles upward from a bedding-plane decollement and begins to step up through the overlying section. The fold develops in front of the fault tip as it propagates. The layers in the upper sheet bend at axial planes (dotted lines). (a) Fault tip in lower dark layer. (b) Fault tip has advanced into white layer. (c) Fault tip has advanced into upper dark layer.

fold, a simple fault-propagation fold, and a fault-bend-fold structure with two imbrications.

Simple fault-bend fold

Fault-bend folding occurs where fault blocks bend as they ride over nonplanar fault surfaces; for example, where a thrust fault steps up from lower to higher decollement horizons. Figure 1 shows three progressive stages in the development of a simple fault-bend fold.

To compute the seismic response to the geologic model in Figure 1c, seismic velocities were assigned to the model which are consistent with an analogue real structure (Figure 2a). As the analogue structure developed concurrent with sedimentation, material points have not been uplifted and are thus at their maximum depth of burial. Because compaction and seismic velocity are controlled principally by depth of burial, velocities will vary *both* with depth and

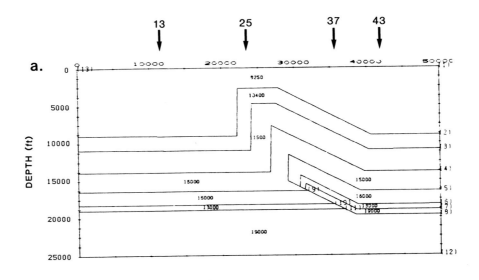

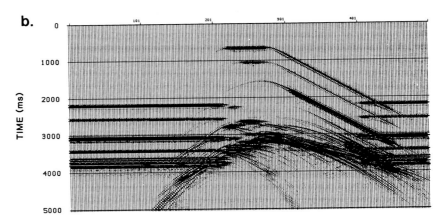

Fig. 5. (a) Velocity model assigned to the final structural model of Figure 4c. Differential compaction effects are not included, so velocity is constant within each layer. Arrows mark the locations of four shot records displayed in Figure 11. (b) Synthetic zero-offset section computed by an exploding-reflector wave-equation algorithm. Picking fault locations and interpreting the core of the fold are problematic.

along stratigraphic horizons. The result is limited horizontal velocity variation (Figure 2a).

This velocity field is not complicated and the resulting synthetic zero-offset section (Figure 2b) is easy to interpret. Since there is little lateral velocity variation, the unmigrated time section appears sufficient for making a correct interpretation. Figure 3a shows that time-migration assists interpretation more by focusing diffractions than by repositioning reflectors. The close match between the time migration and the time model (Figure 3b) confirms that time migration yields an undistorted image for this geologic model.

Simple fault-propagation fold

Fault-propagation folds occur in advance of propagating fault surfaces, such as where a thrust fault along a bedding-plane decollement begins to step up through the overlying section. Figure 4 shows three stages in the development of a simple fault-propagation fold.

To forward-model the seismic response to the late-stage geologic model (Figure 4c), we assign velocities (Figure 5a) that vary only with lithology (i.e., we treat the

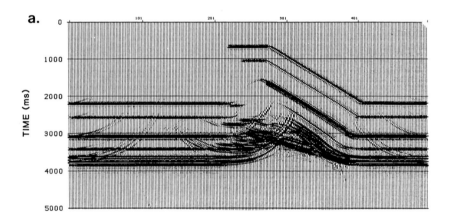

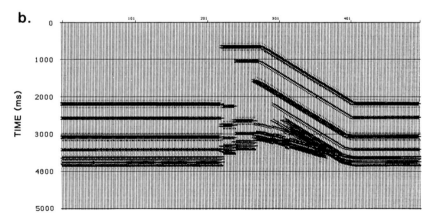

Fig. 6. (a) Time migration of the zero-offset section (Figure 5b) for the fault-propagation fold model. (b) Time model (depth model from Figure 5a converted to 2-way vertical time). Note the differences between the time migration and the time model, suggesting problems in imaging fault-propagation fold structures using time migration.

effects of differential compaction as being insignificant and therefore make velocity constant within each layer).

The resulting zero-offset section (Figure 5b) initially seems relatively easy to interpret for the shallower reflections; however, picking fault locations and interpreting the core of the fold are problematic. Because of the large lateral velocity change, there is a large amount of reflection-time pull-up. Comparing the time migration (Figure 6a) with the time model (Figure 6b) shows that time migration images the shallow events and the right flank of the structure, but fails to image the core where the lateral velocity variation is most rapid. This failure

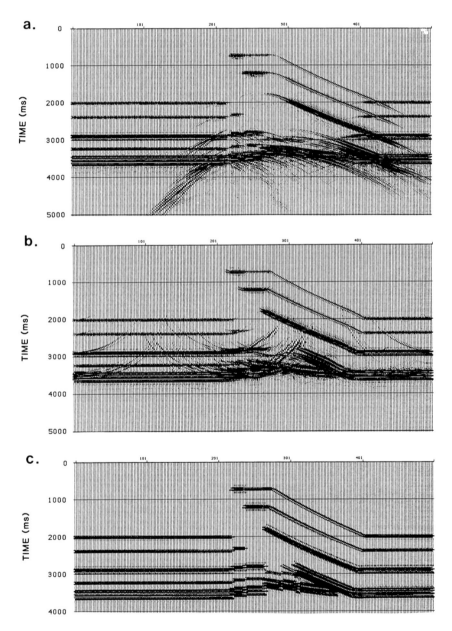

Fig. 7. (a) Synthetic zero-offset section corresponding to the fault-propagation fold model shown in Figure 4c, but computed this time using velocities that vary with lithology *and* depth (compaction). Note the reduced pull-up beneath the structure compared with Figure 5b. (b) Time migration of the zero-offset section. (c) Time model (corresponding to the velocity model that varies with lithology and depth). Discrepancies between the time migration and the time model still exist, but are less pronounced than for the constant-velocity layer case.

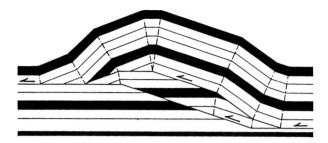

Fig. 8. Fault-bend fold with two imbrications (from Suppe, 1983). Slip along the lower thrust fault imposes additional folding on the overlying older thrust sheet. The presence of multiple imbrications greatly complicates actual fault-bend folds and is commonly observed.

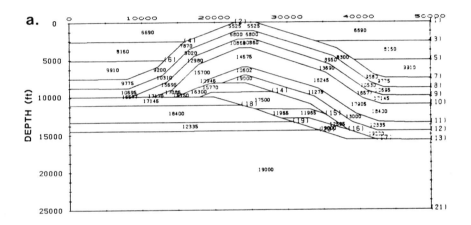

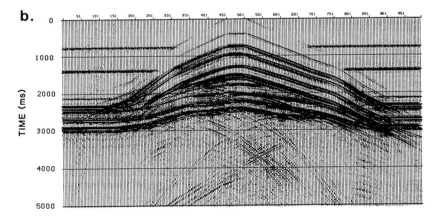

Fig. 9. (a) Velocity model assigned to the imbricated fault-bend fold model in Figure 8. Velocity varies with both lithology and depth (compaction). (b) Synthetic zero-offset section. The presence of an additional imbrication makes this section more complicated than the section in Figure 2b. Real-world complications, such as noise and surface statics, would make interpretation even less straightforward.

suggests that depth migration is probably required in order to obtain a reasonably accurate subsurface image for such structures. Note the absence of reflections from the steeply dipping fore-limb because of finite recording aperture.

To illustrate the importance of compaction effects in the seismic response, the same depth model (Figure 4c) is modeled again using variable velocity layers. In the new model, the velocity of each layer is greater where that layer is buried deeper, simulating compaction. The resulting synthetic (Figure 7a) shows much less reflection-time pull-up than the previous synthetic (Figure 5b). Thus, interpretation of pull-up as being caused by structure or by velocity variation requires (1) knowing velocity variations across a structure from actual measurements or (2) being able to infer velocity variations sufficiently accurately from geologic control (i.e., knowledge of the composition of the rocks and their compaction history). Note that including compaction effects in this model makes lateral velocity variations less severe. Consequently, time migration performs better and the resulting time-migrated section (Figure 7b) yields a less ambiguous interpretation. The time migration agrees fairly well with the corresponding time model (Figure 7c).

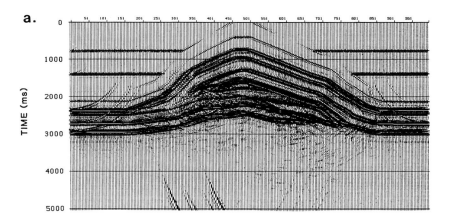

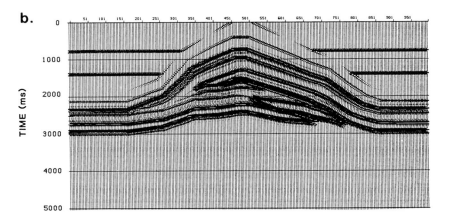

Fig. 10. (a) Time migration of the zero-offset section (Figure 9b) for the imbricated fault-bend fold model. (b) Time model (depth model from Figure 9a converted to two-way vertical time), showing that time migration correctly positioned the reflectors.

Imbricated fault-bend fold

Figure 8 shows a fault-bend fold with two imbrications associated with two episodes of faulting. The presence of multiple imbrications usually complicates interpretation of fault-bend fold structures because the deeper thrusts impose additional folding on the overlying imbricate packages. To forward-model the seismic response to this geologic model, we assign velocities that depend on both lithology and depth (Figure 9a).

Although the structure is idealized and the geologic dips are restricted to 36 degrees or less, the resulting zero-offset section (Figure 9b) is more difficult to interpret (without knowledge of the geologic model) than the single-imbrication synthetic (Figure 2b) discussed earlier. With an irregular recording surface, noise, and other distortions common to real seismic data, the interpretation would be even less straightforward. However, comparing the time migration (Figure 10a) and the time model (Figure 10b) shows that time migration correctly positions the reflections.

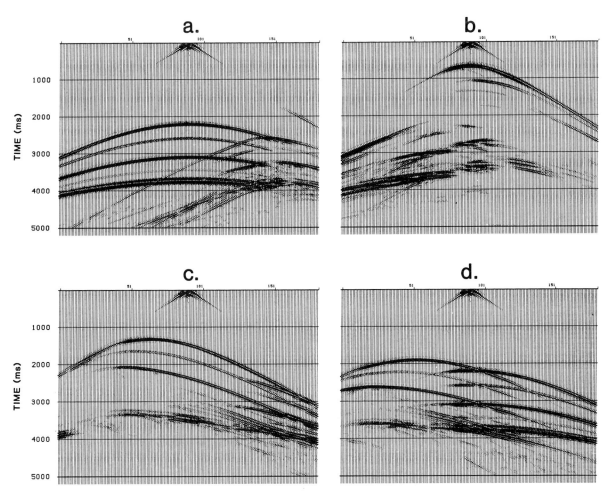

Fig. 11. (a)–(d) Four representative shot records (at the locations marked with arrows in Figure 5a) from a synthetic 2-D seismic survey across the fault-propagation fold model with constant-velocity layers. (a) Shot 13, located at $x = 12\ 000$ ft. (b) Shot 25, located at $x = 24\ 000$ ft. (c) Shot 37, located at $x = 36\ 000$ ft. (d) Shot 43, located at $x = 42\ 000$ ft. A total of 49 shot records were computed across the model using the two-way 2-D acoustic wave equation. The receiver group spacing was 200 ft (61 m), the shot spacing was 1000 ft (305 m), and the maximum offset was 18 000 ft (5486 m).

2-D ACOUSTIC SHOT-RECORD MODELING

A CMP-stack section is commonly treated as if it were a zero-offset section. If multiples are ignored, this assumption is valid for flat layers, but breaks down for complex structures. For example, two events with conflicting dip in a CMP gather may have the same zero-offset time, but will have different stacking velocities. Both events cannot properly be stacked and one will be suppressed in favor of the other. On a zero-offset section, however, both events will be present.

When such differences can significantly affect a structural interpretation, using an approach in seismic modeling that more realistically simulates CMP-stacked data may be advantageous. For this section, we use a two-way acoustic wave-equation algorithm to synthesize multioffset shot records which we CMP gather, NMO correct, and stack. For comparison with the CMP-stack section, we also include the corresponding synthetic zero-offset section and a simple normal-incidence ray-trace section. Differences between the time and depth migrations of the CMP-stack and zero-offset sections help reveal the differences in the information content of the two.

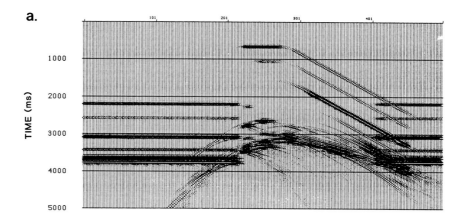

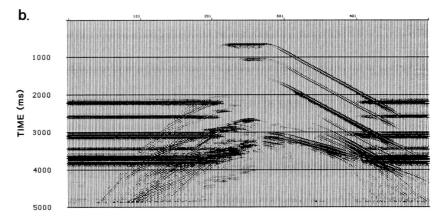

Fig. 12. (a) Exploding-reflector synthetic zero-offset section for the fault-propagation fold model with constant-velocity layers (Figure 5a). (b) CMP stack of the 49 synthetic shot records across the same model. In general, the two appear similar, but reflections from beneath the crest are easier to interpret on the zero-offset section.

Synthetic shot records

We generated shot records every 1000 ft (305 m) across the fault-propagation fold model with constant-velocity layers (Figure 5a). A smaller, more realistic, shot interval would involve an excessive amount of computer time. However, since our synthetic data do not include random noise, a high CMP fold to statistically suppress such noise is not necessary. The simulated survey has a symmetric split-spread recording geometry with 181 traces per shot, a receiver interval of 200 ft (61 m) and a maximum offset of 18 000 ft (5486 m). Four representative shot records (out of the total of 49) are shown in Figure 11; they are from the locations marked with arrows in Figure 5a.

CMP-stack section versus zero-offset section

Figure 12a displays a synthetic zero-offset section, computed for the same model using an exploding-reflector algorithm, but using a different wavelet from that used on the section displayed in Figure 5b. It approximates a perfect, multiple-free CMP stack containing only normal-incidence reflections. Figure 12b shows a CMP stack of the 49 synthetic shot records, which contains multiples and contributions from other than normal-incidence reflections. In general, the two

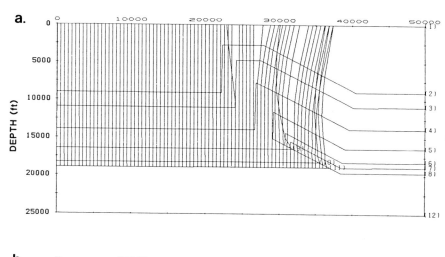

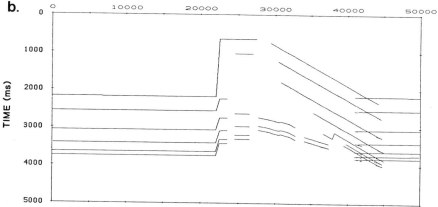

Fig. 13. (a) Normal-incidence reflection raypaths from the deepest subthrust interface, showing this surface can be illuminated from the left all the way up to its contact with the fault. (b) Reflection events predicted by normal-incidence raytracing for all interfaces. Events from the deepest interfaces are discontinuous and have lateral gaps because of the sharp lateral velocity changes in the model.

sections are similar. In generating the zero-offset section, an extracted zero-offset reflection wavelet from the shot-record data was used in order to ensure a more valid comparison with the CMP-stack section. A broader-band wavelet not only would have increased event resolution, but also would have required considerably more computation time for the shot-record modeling program. Reflection events from beneath the crest of the structure appear clearer on the zero-offset section than on the CMP stack. This result is expected because CMP gathers across the crest of the structure are less likely to contain events from common reflection points on all the traces. Another difference is the presence of multiples in the CMP stack which would be more visible with a higher display gain.

Because of rapid lateral velocity variations, one could question whether all of the subsurface can be illuminated and subsequently imaged. Although this question might be answered by migrating the data, ray tracing is a quicker and less expensive method to demonstrate what portions of the subsurface can be illuminated. For example, Figure 13a shows that for normal-incidence ray paths, the deepest subthrust interface can be illuminated from the left all the way up to where the subthrust contacts the fault. However, the reflected waves produced by that illumination do not appear to uniformly reach all parts of the recording surface. The sharp lateral velocity changes in the model result in the reflection

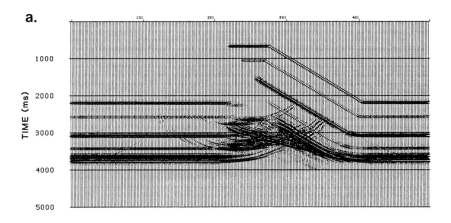

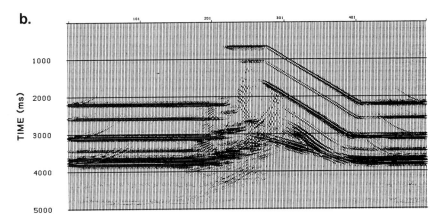

Fig. 14. (a) Time migration of the synthetic zero-offset section (Figure 12a) for the fault-propagation fold model. (b) Time migration of the CMP stack (Figure 12b) for the same model. Similarities between the two sections suggests that the poor imaging under the crest is not because of differences between exploding-reflector and CMP-stack sections.

events predicted by ray tracing (Figure 13b) being discontinuous and having occasional lateral gaps because of low raypath density caused by defocusing. This defocusing suggests that, even when the subsurface is adequately illuminated, to interpret the core of such a structure from an unmigrated section might be difficult. Finally, another advantage of ray tracing, which was useful in this exercise, is the greater ease in matching events in the synthetic section with reflectors in the depth model.

Time migration

The post-stack time migrations (Figure 14) of the zero-offset and CMP-stack sections in Figure 12 illustrate the problems for time migration where lateral velocity variation is significant. These occur even though we used the actual velocity model in the migration. The similarity of the two migrated sections suggests that the poor result is not caused by differences between exploding-reflector zero-offset sections and CMP-stack sections.

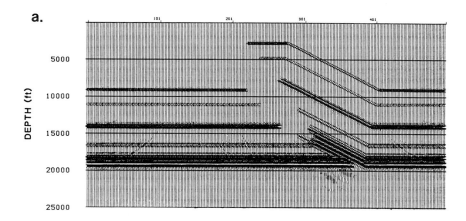

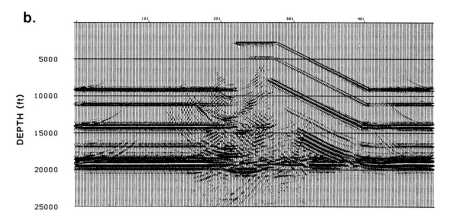

Fig. 15. (a) Depth migration of the synthetic zero-offset section (Figure 12a) for the fault-propagation fold model. (b) Depth migration of the CMP stack (Figure 12b) for the same model. Comparison of these two depth migrations shows that, although depth migration can correctly image the core of the structure, limitations of the CMP method reduce the effectiveness when applied post-stack.

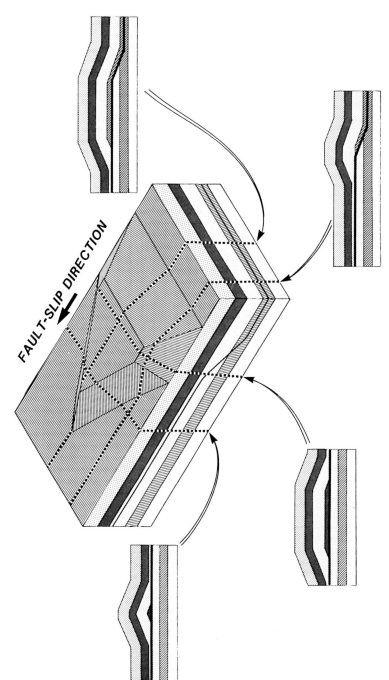

Fig. 16. 3-D fault-bend fold model. The amount of slip is greatest in the vertical plane bisecting the model, is zero at the edges of the model, and varies linearly for parallel planes in-between. Note the similarities between the two cross sections shown for the fault-slip direction and the 2-D fault-bend fold models in Figure 1.

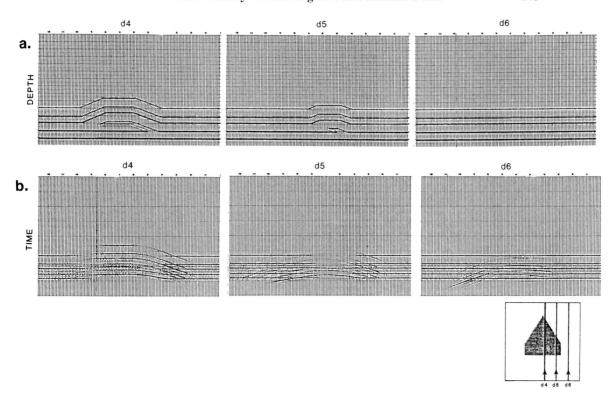

Fig. 17. (a) Slices of the 3-D depth model in the fault-slip direction. The darkened polygon indicates the areas of non-zero slip. (b) Corresponding slices of the 3-D volume of zero-offset (time) traces.

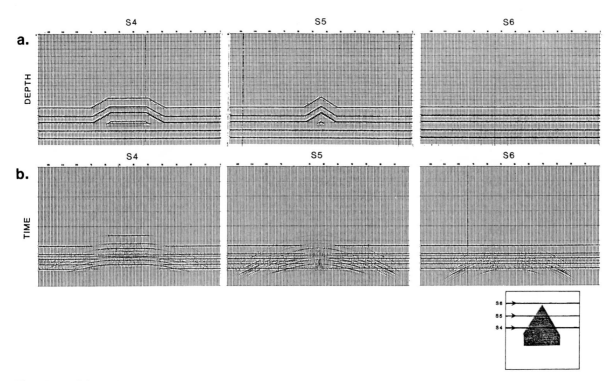

Fig. 18. (a) Slices of the 3-D depth model perpendicular to the fault-slip direction. (b) Corresponding slices of the 3-D volume of zero-offset (time) traces.

Depth migration

The post-stack depth migration of the zero-offset section (Figure 15a) indicates that the structure can be correctly imaged when the input section (e.g., a CMP-stack section) closely mimics a zero-range section and the velocity model used in migration is accurate. (In this example, the migration algorithm is essentially the same as the forward-modeling algorithm used to create the zero-offset section, except that the migration algorithm runs time in reverse.) The depth migration of the actual CMP stack (Figure 15b) shows minimal improvement over the corresponding time migration even though the depth migration accommodates lateral velocity variations. Although depth migration can potentially properly image the core of the structure, the result can still suffer from limitations of the CMP method. Prestack depth migration is a possible way to circumvent such limitations.

3-D FAULT-BEND FOLD MODEL

Now consider a 3-D fault-bend fold structure which can be visualized as a set of parallel 2-D models that are identical in character, except for the amount of slip along the fault (Figure 16). The amount of slip is greatest in the vertical plane bisecting the model, zero at the edges of the model, and varies linearly for parallel planes in-between. As a further aid to visualizing the model, Figure 16 includes vertical slices of the model perpendicular to the fault-slip direction.

For this depth model, a 3-D volume of zero-offset traces was computed using a 3-D exploding-reflector algorithm based on the acoustic wave equation. Figures 17

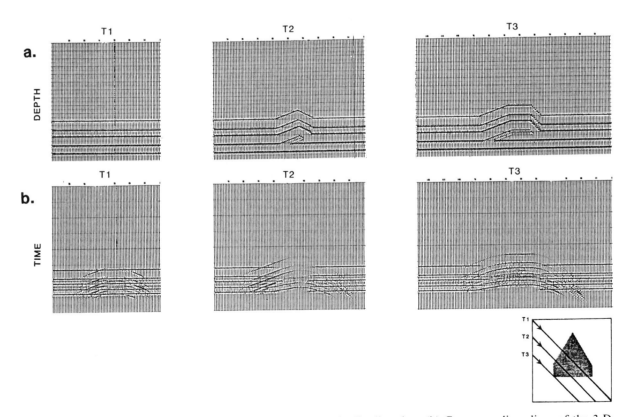

Fig. 19. (a) Slices of the 3-D depth model oblique to the fault-slip direction. (b) Corresponding slices of the 3-D volume of zero-offset (time) traces.

through 19 display various vertical slices of the depth model and corresponding slices of the synthetic seismic data volume. The data slices represent (approximately) CMP stacks that would result from processing data acquired from 2-D seismic lines. Clearly, migration is needed in order to focus the events. Would 2-D migration be sufficient or do we need full 3-D migration?

In order to answer this question, the entire 3-D data volume was 3-D time migrated. Figure 20 displays two parallel vertical slices through each of (a) the depth model, (b) the unmigrated data volume, and (c) the 3-D time-migrated data volume. Part (d) shows the data in part (b) after 2-D time migration. Line X4, which is located off the flank of the structure, is correctly imaged by the 3-D time

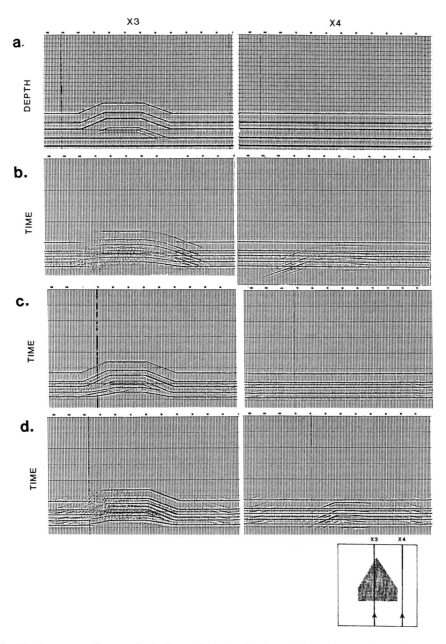

Fig. 20. Corresponding vertical slices of (a) the depth model, (b) the unmigrated synthetic data volume, (c) the 3-D time-migrated data volume, and (d) the 2-D time migration of the sections shown in part (b). 3-D migration provides correct images, while 2-D migration fails because the fault-bend fold is inherently 3-D.

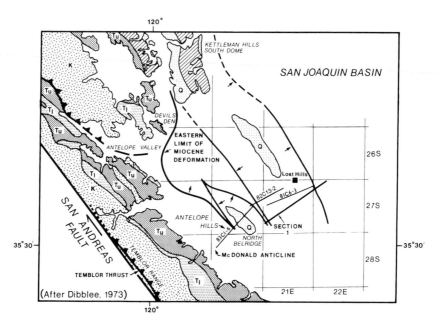

Fig. 21. The Temblor fold belt is located on the eastern side of the San Andreas fault, between the fault and the San Joaquin Basin. East of the Temblor Range, structures in the fold belt are buried by Plio-Pleistocene syntectonic sedimentation. Heavy lines with single arrows indicate the location of axial surfaces of subsurface folds. The location of cross section 1 and seismic lines 81C6-3 and 82C13-2 are annotated. Grid squares are 6 miles (9.7 km) across.

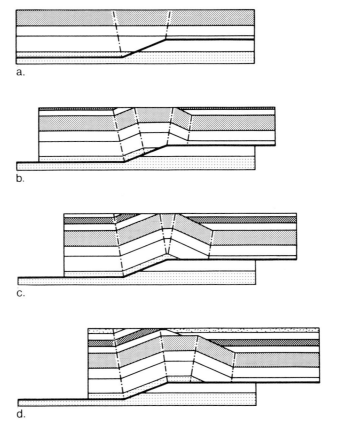

Fig. 22. Theoretical models showing the sequential development of a growth fault-bend fold. The model depicts the case for which the sedimentation rate in the adjoining basin exactly equals the instantaneous uplift rate above the ramp. Note the apparent normal-fault offset along the disconformity on the crest of the anticline in part (d).

SOUTHEAST LOST HILLS

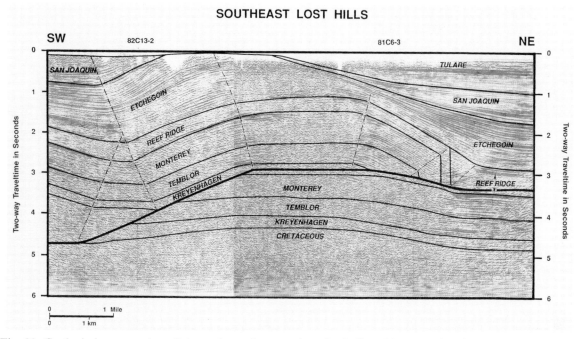

Fig. 23. Geologic interpretation of time-migrated composite seismic line. The unconformity surface which begins within the Etchegoin Formation on the northeast side of the anticline rises stratigraphically over the anticline, ending up at the top of the San Joaquin Formation on the southwest side of the fold.

SE LOST HILLS

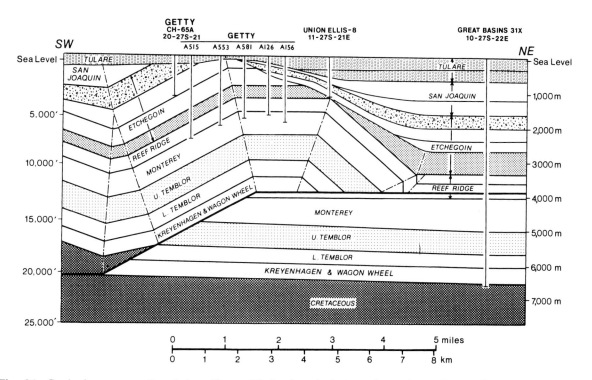

Fig. 24. Geologic cross-section 1 (see Figure 21 for location) through southeast Lost Hills showing growth fault-bend fold solution. Time-transgressive unconformity surface begins within the Etchegoin Formation on the northeast side of the anticline and rises stratigraphically over the anticline, ending up at the top of the San Joaquin Formation on the southwest side of the fold.

migration (Figure 20c, right panel), but is incorrectly imaged by 2-D time migration (Figure 20d, right panel) which shows events caused by sideswipe. Because line X2 is located nearly axially on the structure, there are fewer differences between the 3-D and 2-D time migrations (Figures 20c and 20d, left panel, respectively). Still, the 3-D migration of line X2 gives a better time image of the corresponding depth slice.

The 2-D migration is limited since reflections are assumed to come from within the vertical plane. 3-D methods (acquisition and processing) must therefore be used to image 3-D structures. With a priori dip information on selected target reflectors, and by using 2-D methods, sometimes the seismic data acquisition and processing can be arranged in order to get a reasonably accurate image of the target.

REAL DATA EXAMPLE

2-D seismic modeling can be used in conjunction with geologic constraints to generate an improved velocity model from an initial velocity model. This is part of a case study (Medwedeff, 1989) of the Lost Hills anticline, which is located

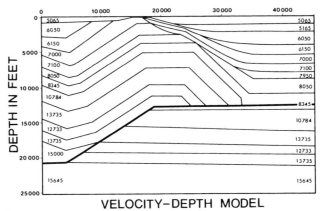

a.

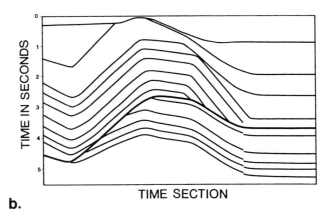

b.

Fig. 25. (a) Velocity model (Model 1) for Lost Hills anticline assuming constant-velocity layers. (b) Time model corresponding to Model 1. Note the exaggeration of limb dips in the time section.

along the west side of the San Joaquin Valley, California (Figure 21). The anticline is upright, slightly asymmetric, 6 mi (10 km) wide and 20 mi (32 km) long. Fold growth during the Late Miocene through Quaternary time is indicated by onlapping sediments of the same age.

The dominant features of the Lost Hills structure can be explained using growth fault-bend fold theory (Medwedeff and Suppe, 1986). Figure 22 displays the sequential development of a fault-bend fold in which the rate of uplift of the anticline equals the sedimentation rate in the adjoining basins. The striking similarity between the Lost Hills anticline, as observed in seismic data (Figure 23), and the growth fault-bend fold model (Figure 22d) suggests that the Lost Hills anticline is a northeast-verging growth fault-bend fold.

High-quality borehole information and time-migrated seismic data (Figure 23), together with fault-bend fold theory, are used to construct a retrodeformable geologic section to a depth of 25 000 ft (7622 m) (Figure 24). Using sonic-log velocities from the Great Basins 31x well in the San Joaquin Basin, a velocity field is assigned to the geologic model (Figure 25a). This initial velocity-depth model assumes that velocities within each layer are constant. The corresponding time

SOUTHEAST LOST HILLS
MODEL 2: VARIABLE VELOCITY LAYERS

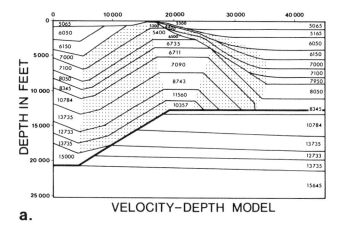

a.

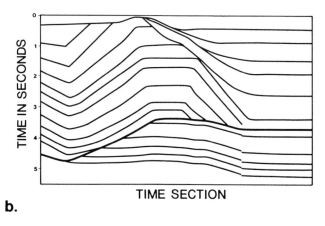

b.

Fig. 26. (a) Velocity model (Model 2) for Lost Hills anticline using a three-part velocity scheme. The crest of the anticline is assigned one velocity profile, the flanking basins another, and linear interpolation is used to specify the velocities of the fold limbs (stippled area). (b) Time model corresponding to Model 2. Note the relative lack of dip distortion and pull-up.

model (Figure 25b) exhibits steeper-dipping limbs and more pull-up of the subthrust units than seen on the time-migrated section from Lost Hills (Figure 23).

Rather than making velocities constant within layers, a more realistic description is needed. Seismic velocity in a clastic section is a function of compaction and, therefore, of maximum depth of burial (Sheriff and Geldart, 1983, p. 7); hence, the seismic velocity of rock at any given point in the structure depends on the rock's particular burial history as well as on the lithology. To properly modify the velocity-depth model, the burial history of the stratigraphic units must be reconstructed. Inspection of the sequential development of a growth fault-bend fold (Figure 22) indicates stratigraphic units move down, or are stationary, with respect to the ground surface. Thus, at any time in the development of the structure, all points in both the pre-growth and growth units are at their maximum

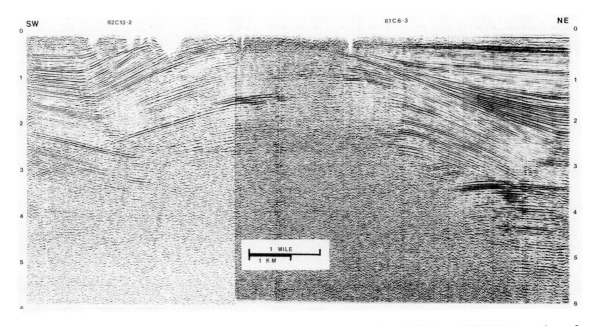

Fig. 27. CMP-stack unmigrated section for composite seismic line (from lines 81C6-3 and 82C13-2) at southeast Lost Hills.

ZERO-RANGE SYNTHETIC SEISMIC TIME SECTION OF VELOCITY-DEPTH MODEL 2

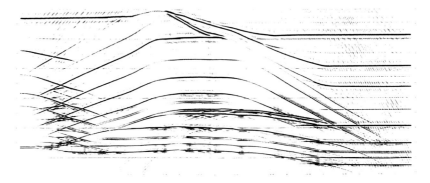

Fig. 28. Exploding-reflector synthetic zero-offset time section for Model 2 (Figure 26a). Note the overall similarity with the CMP-stack section (Figure 27).

depth of burial. Therefore, although the thrust fault duplicates stratigraphic section, the fault does not duplicate the seismic velocity field.

In the modified model (Figure 26a), the velocities are assigned differently to three distinct parts of the geologic section. The basin areas on either side of the Lost Hills anticline are assigned average formation velocities calculated from a sonic log of the Great Basins 31x well. The relatively flat crest of the anticline is given average formation velocities from the Mobil Williams 33 well located on the anticline. The velocity structure is completed within each layer by linear interpolation down the fold limbs. Although velocity-depth relationships are not linear for rocks (Sheriff and Geldart, 1983, p. 7), this simplification provides an approximate model in which the velocity of each layer increases with depth. The time model displays limited distortion of dips and pull-up of the subthrust units and more closely resembles the time-migrated section of Lost Hills (Figure 23) than does the earlier synthetic.

To further check the consistency of the modified model with the seismic data from Lost Hills, an exploding-reflector synthetic zero-offset seismic section was generated (Figure 28). Qualitative similarities between the final stacked unmigrated section (Figure 27) and the synthetic section (Figure 28) suggest that the geologic model provides a valid interpretation.

CONCLUSIONS

Processing the exploding-reflector synthetic zero-offset sections reveals that post-stack time migration is generally adequate for most simple fault-bend fold geometries. Depth migration may often be required for fault-propagation folds, which usually involve more rapid lateral velocity changes than do fault-bend folds. Depending on lithology and burial history, the effects of compaction on seismic velocity can be significant in these structures; they should be taken into account in data processing, especially if depth migration is to be applied.

Comparing the CMP stack of synthetic shot records across a fault-propagation fold with the corresponding exploding-reflector zero-offset synthetic indicates that CMP stack has been somewhat successful in matching the expected zero-offset section. Under the crest of the fold, however, where the common-reflection point assumption weakens, the full CMP stack provides a less-clear picture than the zero-offset data. This observation is supported by visual inspection of the time- and depth-migrated sections. Raytracing indicates that at least some paths exist for waves to travel down to and back from the subthrust reflectors, so those reflectors are being illuminated, despite their weak images on the migrated CMP-stack sections.

Comparing the 2-D and 3-D migrations of the 3-D zero-offset data recorded across a 3-D fault-bend fold shows that 3-D data acquisition and processing are necessary to enable a correct structural interpretation; 2-D lines oriented in the fault-slip direction are insufficient.

The Lost Hills case history shows how zero-offset modeling can be used to iterate toward an internally consistent solution. This is a good example of how iterative combined geologic and seismic modeling can be facilitated by using objective geologic control, such as fault-bend fold theory combined with subsurface well information.

ACKNOWLEDGMENTS

We are indebted to J. H. Higginbotham and D. V. Sukup for providing the wave-equation programs and advice. Special thanks go to G. T. Chou for automating the geologic modeling of fault-related folds, J. M. Evans for assistance in conducting the case history, and C. L. Robertson for providing effective model-digitizing and editing capabilities. We also thank G. S. Edwards and R. H. Tatham for reviewing the manuscript.

REFERENCES

Dibblee, T. W., Jr., 1973, Regional geologic map of the San Andreas and related faults in Carrizo Plain, Temblor, Caliente, and La Panza Ranges and vicinity, California: USGS Misc. Geol. Invest. I-757, scale 1:125000.

Kjartansson, E. and Rocca, F., 1979, The exploding reflector model and lateral variable media: SEP Rep. 16, Stanford Exploration Project, Stanford Univ.

Judson, D., Lin, J., and Sherwood, J., 1980, Depth migration after stack: Geophysics, **45**, 361–375.

Medwedeff, D. A., 1989, Growth fault-bend folding at Southeast Lost Hills, San Joaquin Valley, California: Bull., Am. Assn. Petr. Geol., **73**, 54–67.

Medwedeff, D. A., and Suppe, J., 1986, Growth fault-bend folding—precise determination of kinematics, timing, and rates of folding and faulting from syntectonic sediments: Geol. Soc. Am., Abstracts with Program, **18**, 692.

Sheriff, R. E., and Geldart, L. P., 1983, Exploration seismology, 2, Cambridge Univ. Press.

Suppe, J., 1983, Geometry and kinematics of fault-bend folding: Am. J. Sci., **283**, 684–721.

——, 1985, Principles of Structural Geology, Prentice-Hall, Inc.

Suppe, J., and Medwedeff, D. A., 1984, Fault-propagation folding: Abstracts with program, Geol. Soc. Am., **16**, 670.

CASE HISTORY 5

Ray-Trace Modeling for Salt Proximity Surveys

E. A. Nosal, C. J. Callahan* and R. W. Wiley‡*

ABSTRACT

A Gulf Coast salt dome provides the setting for a case history using seismic ray trace modeling in the design and interpretation of a salt proximity survey. The salt proximity survey was undertaken in order to have an independent estimate of the position of the salt flank because an exploration well had penetrated porous sands but surface seismic of two different vintages gave conflicting interpretations of the amount of salt overhang. If sufficient salt overhang existed then a sidetrack well could be justified to test updip extension of the sand.

The salt proximity survey had two parts, one part to image the salt flank at shallow depths and the other to image deeper. Two computer models were constructed corresponding to the two surveys. Various shot locations were tested on the models for ascertaining the zone of salt imaging and associated time-depth curves. The graphic displays of ray patterns from the ray-tracing computer program can be used to develop quality control check routines useful in the field during the acquisition phase of the project. These routines show how some potential interpretative problems can be resolved by the example of inserting a caprock on one of the models to study the change of salt face imaging that occurs. An aplanatic analysis of many ray tracing trials were made in order to understand the effect of parameter changes on the subsequent analysis.

Finally, results of the field program are discussed. Results show that a salt proximity program can be successfully run and integrated with surface seismic information. A large part of the success of the survey was due to the presurvey planning and preparedness that came from the modeling process.

INTRODUCTION

Marathon Oil Company is active in the U.S. Gulf Coast region where important oil and gas accumulations are found around salt domes. Often, the reservoir sands dip upward toward the diapir and the hydrocarbon accumulations are trapped by the impermeable salt flank. The sides of a salt diapir are often shaped irregularly, and can have undulations and overhangs which make seismic interpretation difficult at times. When interpreting the position of a salt face requires more accuracy than can be obtained from seismic, and when a nearby well is available,

*Marathon Oil Co., P.O. Box 3128, Houston, TX 77253.
‡Marathon Oil Co., Exploration & Production Technology Center, P.O. Box 269, Littleton, CO 80122.

Marathon often uses salt proximity surveys for additional, independent information.

The presented case study illustrates incorporation of ray trace modeling into the acquisition and interpretation stages of a salt proximity survey. Modeling significantly improved the quality of the program. Modeling was used to initially plan the survey and later to examine alternative interpretations of the field data. Another benefit from modeling came through the process of modeling itself because creating a model required careful, iterative analysis of the geologic data and geologic hypotheses. Armed with knowledge gained through modeling studies better examination of contingent situations in program design and final evaluation were possible.

The ray trace modeling was done to investigate two key issues: (1) to determine the portion of the salt face that would be imaged for a potential shot point and (2) model variations were generated to assist us in interpreting the field data we ultimately acquired. In this regard, modeling served to constrain the possible solutions of the aplanatic analysis by helping us put bounds on the effects of several poorly known factors (for example, the sediment velocity above the top of salt). The graphic output of the modeling calculations are: (1) plots of ray paths from shots over the salt into receivers along the well bore, and (2) the associated first break time-depth recordings. These output were used to understand the physics of each geologic alternative and were compared with the acquired data to guide the final interpretation.

First, the nature of the seismic problem investigated is discussed. Then a short explanation of salt proximity surveys and aplanatic analysis is provided. Finally, results of this study are presented. We clearly show that ray trace models served as a vehicle for carrying hypotheses about the geology from the initial stages of the salt proximity program to the final interpretation.

SALT DOME SETTING

The setting for the salt proximity survey is in the offshore Gulf Coast. A top of salt time structure map is shown in Figure 1. The map was constructed from seismic and shallow hazard survey data. A vertical well (designated MOC) was drilled at the flank of this salt dome and porous sands were encountered at the target depths. Because the penetrated sands were wet, a sidetrack well would normally be warranted to test the updip extension of the sands. However, doubt existed as to whether there was sufficient updip potential because two different vintages of CMP seismic data supported two different interpretations of the salt face location.

Two seismic sections shown in Figures 2a and 2b illustrate the problem. The lines are shown in Figure 1 to cross at the well location. The shaded regions show the initial estimate of the salt face location. The interpretation on Line 1 indicates little salt overhang and, therefore, insufficient updip closure. The interpretation of Line 2, however, made a case for economic updip potential because of the large amount of salt overhang indicated. The conflicting interpretations could not be resolved using seismic data alone, so a salt proximity survey was shot to obtain independent evidence.

SALT PROXIMITY SURVEYS AND APLANATIC ANALYSIS

Early papers, Gardner (1949) and Musgrave et al. (1960), discuss salt proximity surveying and aplanatic analysis. The basic concepts discussed by these authors are incorporated in a Marathon developed computer program which permits entry of a complex velocity model and generates aplanatic curves for a complete suite of recordings. A few of the important points on the subject are reviewed for those who are unfamiliar with this type of survey.

A salt proximity survey is a seismic refraction survey conducted with a source near the surface and with geophones in a borehole. Typically, the source is placed over the top of a salt dome and the chosen borehole is situated near the flank of the dome. With this geometry the seismic ray path enters at the top of the salt dome, passes through a portion of the salt dome, and exits the face of the salt dome in the region to be imaged. The seismic ray path also passes through sediments above the top of salt and through sediments that may lie between the salt face and the geophone. Thus, the total traveltime from source to receiver is composed of four time segments (Figure 3):

(1) from source to water bottom;
(2) from the water bottom to the top of salt (i.e., through sediments above the salt), and through an anhydride cap (if one exits);
(3) from top of salt to the point at the salt face where the energy exits the salt dome;
(4) through the sediments between the borehole and the flank.

The primary data acquired are the total traveltime. The purpose of the analysis is to estimate the portion of traveltime spent in the salt mass. That estimate is then used to compute the position of the salt face.

Consider first a single experiment from one shot into one geophone in the borehole. The aplanatic analysis requires a depth model of all the sediments and their associated velocities along the seismic travel path. In addition, the depth to

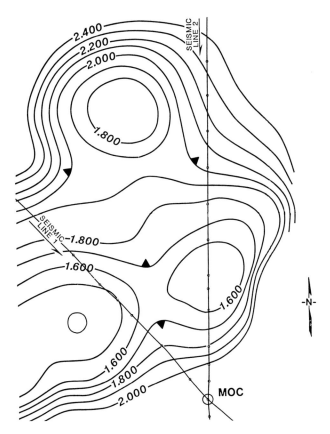

Fig. 1. Top of salt map with location of rig and seismic lines.

the top of salt at the shot location and the salt velocity must be known. Much of this information is acquired from checkshot surveys, log suites, and directional surveys. Therefore, the only unknown parameter is the position of the salt face (or, equivalently, the traveltime spent in the salt dome). Aplanatic analysis proceeds by examining all possible salt face positions (all possible traveltimes in the salt) within the model just described. Each possible salt face position is defined as a position that satisfies seismic propagation within the model and which gives a modeled total traveltime equal to the measured total traveltime. In general there are very many positions which satisfy these constraints. The locus of all possible salt face locations is called the aplanatic curve. Each point along the

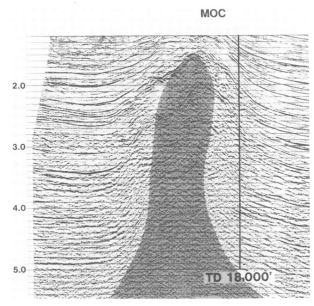

Fig. 2a. Line 1 (vintage 1984) with salt dome interpretation.

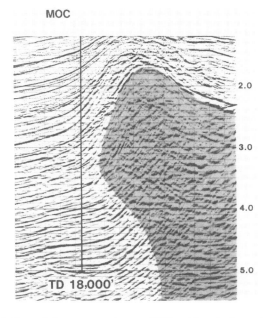

Fig. 2b. Line 2 (from 3-D survey, vintage 1985) with salt dome interpretation.

aplanatic curve is a possible position of the salt face with a traveltime equal to the measured traveltime.

If only a single experiment were conducted then the best that could be achieved is the range of salt face positions inherent in the aplanatic curve. In practice, however, recordings are made with geophones placed at many depths along the wellhole. Each source-receiver pair yields an individual aplanatic curve and the collection of such curves more accurately determines the actual salt location. The aplanatic curves from all the recordings are plotted on the velocity model and the salt location is estimated by an envelope to the family of curves. Examples of such curves are shown in Figures 4d, 11a, and 12a.

CHOOSING A SOURCE LOCATION

The objective of the salt proximity survey was to determine the position of the salt overhang where the potential reservoir sand was truncated. Therefore, this area of the salt face had to be seismically illuminated. Ray trace modeling is a powerful design tool which helped ensure that this acquisition requirement was achieved. This section of the case study describes some of the ray tracing trials we undertook as a prelude to the acquisition program.

We used a vendor program that computes reflected and refracted ray paths for *P*- and *S*-waves including mode conversions, but not critically refracted energy. The program runs on our Digital VAX computers. The speed of the computations is such that model building and ray computations are interactively performed. This convenience allows easy changes and iterations to the velocity model and any other important parameter variations.

Two models produced in the course of the study were used for defining survey geometry. Model 1 (Figure 4a) was created to intersect the well and run between the two seismic lines shown in Figure 2. This model represents our best understanding of the geology at the time the model was built. The salt dome lies on the left side of the model and is given a seismic velocity of 14.7 kft/s (4480 m/s). The seven sedimentary layers we used in the model were interpreted from well logs and seismic data, and their associated velocities were derived from check shot data and velocity logs. The dip angles of the layers near the salt flank are approximately 20 degrees. Figure 4a also shows three potential shot locations for

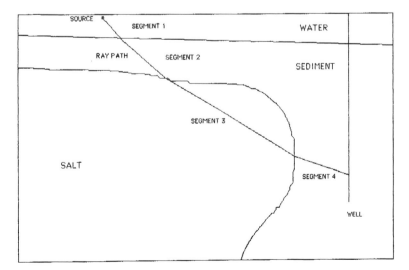

Fig. 3. Diagram of the major segments of a salt proximity ray path.

the salt proximity survey, labeled at the top. Our task was to find a shot position that would image the salt face as low as possible along the overhang.

A part of the array design in the modeling program allows a choice of geophone spacing in the wellbore. The chosen sampling interval is based on a simple Nyquist calculation. A maximum wavenumber equal to 0.005 cycles-per-foot is obtained by using 40 Hz for the maximum source frequency and minimum sediment velocity of 8000 ft/s (2438 m/s). By the Nyquist cutoff formula the spatial sampling interval comes to 100 ft (30.48 m).

From the standpoint of modeling as dense a geophone array as one wishes can be chosen, although there is a computational time penalty for over sampling the wavefront. On the other hand, although one or two rays would seem to be sufficient to provide first arrival information, there is a risk that too sparse a sampling may yield biased results due to local variations in the model. A wave reconstruction criterion is used at this time to determine the geophone spacing because we anticipated that the borehole in the field intercepts the wavefront we wish to record.

Reasons for recording the wavefront in the field are, first, the recorded passage of the wavefront, i.e. the time-depth curve, has a shape influenced by the shapes of the salt flank and the borehole trajectory. Modeling provides a preview of how those shapes might look, and these shapes serve as quality control references during acquisition. The wavefront presumably would heal from any effects of small surface irregularities on the top of the salt dome. Second, a salt proximity survey is a form of VSP and we record the full waveform in the field in order to look for later arriving events, such as water bottom multiples or mode converted energy. Third, the ray paths in the model allow us to project segments of wavefront seen in the field data back to the area of salt face that is imaged. Fourth, accurate picking of first arrival times in a noise filled environment is greatly facilitated when an adequately sampled wavefront is available.

The ray tracing code operates by sending out a fan of trial rays from the source. Then the program searches to find how closely each ray may come to each

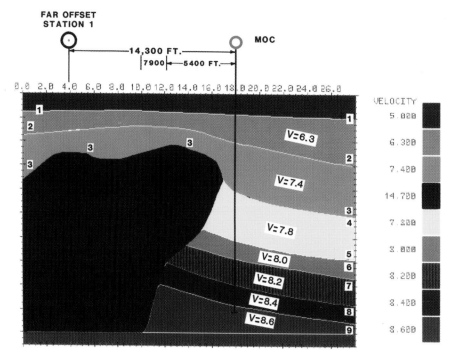

Fig. 4a. Model 1. Station 1 is located 14 300 ft (4359 m) from the well.

geophone. When rays come within a prespecified distance of a geophone (we usually choose half the geophone spacing) they are counted as being detected by the geophone. When a geophone is missed by rays because the ray path passes outside the prespecified detection circle, the program performs iterations on internal ray generation parameters and attempts to calculate a ray that will meet the detection criterion. If the iteration algorithm cannot converge, that geophone will not record an arrival.

The first trial position was at 5400 ft (1646 m) from the well head, almost directly over a salt crest (Figure 5b). This position will serve to demonstrate some of the analyses that can be performed on the graphic products of the ray tracing program. A diagram of the ray path distribution is shown in Figure 4b. Because the critical angle of refraction is 30 degrees for the velocities in this model, the entry angles of the directly transmitted P-wave rays will be those concentrated around the normal vector to the surface. Figure 4b shows that the direct arrival rays which energize the geophones do enter the salt top in a narrow bundle, almost directly beneath the surface location. The most important observation in this diagram is the area of salt face that is imaged. The fan of rays exit the salt in a zone from 7000 to 12 000 ft (2134–3658 m) TVD, which lies above the geophones that record the event. Observe also that all the rays entering the salt are concentrated in a very small area on the top of the salt. As a result of the small entry area, the ray pattern within the salt mass looks as if a pseudosource was located on the top of the salt and is illuminating the salt flank.

The associated time-depth panel, shown in Figure 4c, is a spike section. In order to simplify the time picking the section has not been convolved with any wavelet. Each vertical line is a geophone trace positioned at 100 ft (30 m) intervals down the well. Notice that the velocity contrasts of the sediment layers are small enough that the event of first arrivals, which we call a moveout curve, traces out a smooth, almost straight line. Also, only a few of the geophones have failed to capture rays from this shot position. The slope ($\Delta t/\Delta x$) to the time-depth curve yields a reciprocal parameter we call the moveout velocity ($\Delta x/\Delta t$), and is

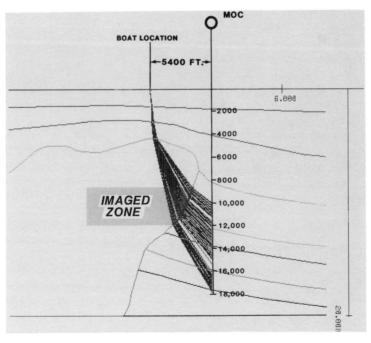

Fig. 4b. Ray paths for Model 1 which show 5400 ft offset. Salt face imaged between 9000 to 12 000 ft TVD.

measured from this diagram to be about 11 250 ft/s (3429 m/s). This value is a consequence of the high salt velocity plus the highly acute angle at which the rays intersect the borehole. A graphic measurement on Figure 4b gives average angles of incidence of about 45 degrees from vertical for the rays in the sediment layers next to the borehole and about 25 degrees from vertical in the salt mass. For the geophone placed at 16 000 feet (5568 m) TVD the straight line (no refraction at the salt/sediment interface) travel path to the pseudosource location is about 23 degrees from vertical. The dependence of the moveout curve on the model geometry can be understood from considering the case of plane waves in a homogeneous medium. The moveout velocity equals the medium velocity divided by the sine of the angle of emergence (angle of emergence is measured from the ray vector to the perpendicular of the borehole) of the wavefront (Slotnick, 1959). Conversely, we can define an apparent medium velocity in this inhomogeneous medium by multiplying the moveout velocity by sine (90–23 degrees) to obtain 10 350 ft/s (3150 m/s), a higher value than the sediment velocities, and approaching the salt velocity. In a similar manner, the straight line to the pseudo-source location for the geophone placed at 10 000 feet (3048 m) TVD is 41 degrees from vertical and leads to an apparent medium velocity of 8490 ft/s (2590 m/s), a value close to that of the sediments. Both of these calculations of apparent velocity show how we can perform quick quality control checks of the field data as the data is acquired.

Further observation indicates that the average speed of the rays is slightly higher in the upper geophones than in the lower geophones because the upper rays travel proportionately farther in the salt than do the lower rays. We measured the lengths of the ray segments in the salt and in the sediments and calculated an average velocity from these ray lengths and the velocities of the material traversed by each ray segment. The ray that travels from the pseudosource to the geophone in the shallow part of the borehole travels 71 percent of that distance in salt, while those rays in the lower part of the borehole travel 54 percent of their distance in salt. These rays yield average velocities of 11 700 ft/s (3566 m/s) and 11 000 ft/s (3353 m/s), respectively. Thus, the apparent medium velocity increases with depth but the ray average velocity decreases. Computing an apparent velocity offers a

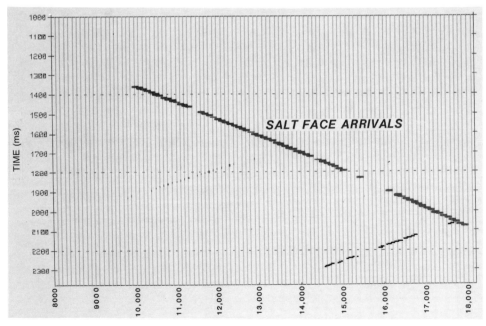

Fig. 4c. Seismic record for Model 1 with shot offset of 5400 ft.

means of checking the data in the field at the time of acquisition. The average velocity is a physically meaningful quantity, but impossible to calculate without detailed knowledge of velocities and ray-travel paths.

Figure 4d shows a graphic output from our Marathon developed aplanatic analysis program. The program was given the same sediment velocity model, top of salt model, and salt velocity that were used in Model 1. The two bounding spatial points indicated on the top of salt are a point directly below the shot and the point $(x_c/2)$ where the rays approximately attain their critical angle. The aplanatic curves are the almost circular curves clustered to the left and at the bottom half of the borehole trajectory trace. An aplanatic curve is computed for every geophone. The interpretation of the salt face position, made by the envelope of tangents to the family of such curves, matches the original model salt geometry in this case. Clearly, having many geophone stations produces a dense family of curves and enables a more reliable interpretation than if only a few recordings were made.

The effect of increasing the shot offset distance to 7900 ft (2408 m) is demonstrated in Figure 5a, where the shot position is above the back side of the nearer salt crest (see Figure 5b). The imaged zone has not changed appreciably, still reaching down to about 12 000 ft (3658 m) TVD.

When the shot position is placed over the second salt crest, Figure 6, at an offset of 14 300 ft (4359 m), salt face illumination reaches as deeply as can practically be expected under field conditions. At this shot position the imaged zone is lowered to close to 13 000 ft (3962 m) TVD. Moving the shot farther away risks the quality of the recording because noise (such as rig noise) will dominate the signal as propagation losses mount. In addition, confidence in the interpretation may decline out of our poorer knowledge of sediment properties farther away from the well location. Again, the concentration of rays around the normal to the top of salt surface appears to indicate that there is a pseudosource located at the ray entry position on the top of salt. Moreover, the arrangement of the pseudosource and the salt flank constrain the ray paths to nearly straight lines between the pseudosource and geophones for all geophone stations. These rays intersect the well at about 47–67 degrees from vertical. A graphic measurement of ray lengths on Figure 6 yields average velocities of 13 000 ft/s at the upper geophones and 11 900 ft/s at the lower geophones. The higher average velocity at the upper geophones occurs because the rays to the upper geophone travel proportionately farther in the salt than do the rays to the lower geophones. The moveout velocity, determined from the time-depth curve (not shown), decreases from about 18 000 ft/s in the upper region to 10 000 ft/s in the lower borehole. This diagram also suggests how each individual aplanatic curve can be interpreted for the salt location when the ray paths between borehole and pseudosource are straight lines. Because the intersection point of the aplanatic curve with the line segment from the geophone to the pseudosource is used as the salt position then the direction to the pseudosource is the same as the direction of wavefront travel. When the spatial sampling is sparse, causing difficulty in drawing an envelope of tangents, then this estimate can be useful.

In the course of examining this model we found an interesting combination of events when the shot was placed at an offset of 14 100 ft (4298 m). This placement of the shot is not much of a change from the previous positions but does serve to alert the interpreter to a possible source of misinterpretation. Figure 7a shows the pattern of directly transmitted rays intersecting only a few geophones at the bottom of the well. This behavior is a consequence of salt top geometry at that particular source location. The tangent to the top of the salt surface, at the point where the ray enters the top of salt, is an upper limit to the angular distribution of rays that can be transmitted into the salt since the tangent defines the direction of critically refracted rays. At the front slope of the salt crest the tangent points

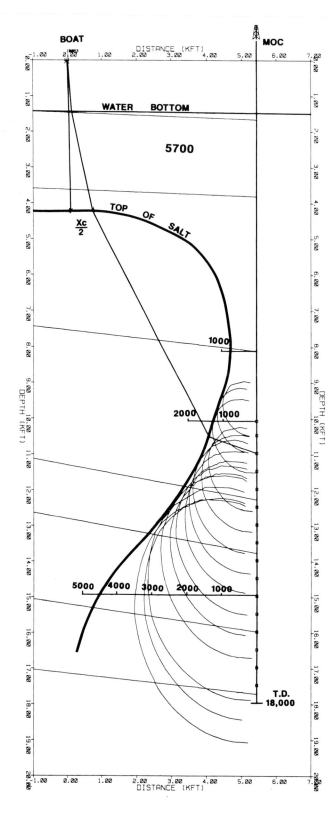

Fig. 4d. Aplanatic curves and interpretation for Model 1 and shot offset of 5400 ft. Trial sediment velocity above salt is 5700 ft/s.

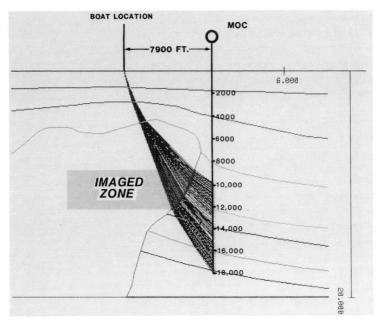

Fig. 5a. Ray paths from Model 1 with shot offset of 7900 ft. Salt face imaged between 9000 to 12 000 ft TVD.

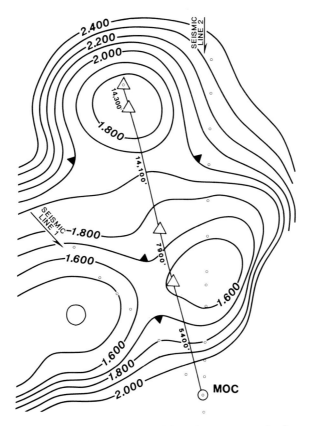

Fig. 5b. Location map of model stations on top of salt map.

downward and intersects the borehole low. Thus, there is a chance that direct, transmitted rays only reach the lower portion of the wellbore when the shot location lies over the front slope of the salt crest. All entry angles greater than 30 degrees will give rise to mode conversions that have different amplitude and phase relationships from the direct *P*-wave arrivals. These events might be difficult to recognize. In Figure 7a we see that the slope of the crest is great enough that only the very lowest geophones in the well are reached by direct arrivals.

In a real experiment there are other ray paths in addition to direct *P*-wave transmission. One such complicating factor is a first-order multiple from the water

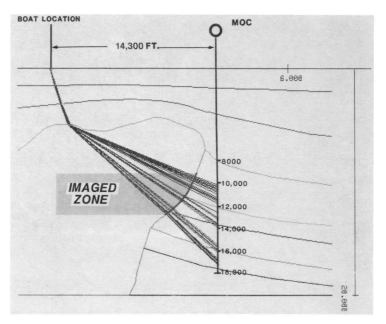

Fig. 6. Ray paths for Model 1 with shot offset of 14 300 ft. Salt face imaged between 10 000 to 13 000 ft TVD.

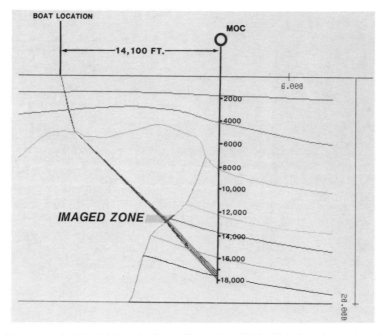

Fig. 7a. Ray paths for Model 1 with shot offset of 14 100 ft. Salt face imaged at 13 000 ft.

bottom. The ray pattern of a water-bottom multiple is shown in Figure 7b for the shot located at 14 100 ft (4298 m). This ray pattern energizes many more geophones along the borehole than did the direct rays and illuminates a broader part of the salt flank as well. Figure 7c shows the time-depth moveout curves for the two travel paths, namely the direct *P*-wave and the water-bottom multiples. We believe that a complex wavefront, with these seismic responses, would likely lead to errors in interpretation because the deep geophones would record strong, continuous events of direct *P*-wave energy while the geophones higher up the borehole would record mode converted energy first and then the strong, contin-

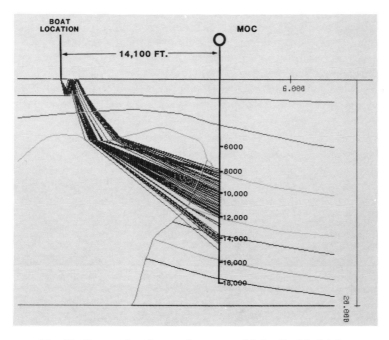

Fig. 7b. Ray paths of water-bottom multiples for Model 1.

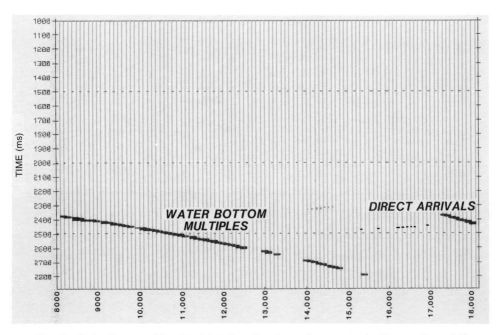

Fig. 7c. Seismic record for Model 1 showing time of ray paths in Figures 7a and 7b.

uous response of the water bottom multiple. In the worst case the time-depth section might look like Figure 7c. The interpreter might mistake the water bottom multiple as his first break time and regard the preceding energy as noise. By using ray trace modeling we were alerted to this problem and avoided the issue by choosing a shot location toward the back side of the crest.

We decided on placing a first source location at 14 300 ft (4359 m) to image the lower part of the wellbore (Figure 6). A second shot position was needed to image the shallow portion of the salt flank.

The second shot location was chosen to be along seismic line 2 (Figure 8). Model 2 was created for this survey, and is shown in Figure 9. At this location there is only a single salt crest. The parameters of the model were, of course, selected on the basis of the best geologic data we had at the time. The model shot location shown in Figure 9, 8500 ft (2591 m) offset from the well, was approximately equal to the distance we ultimately used for the survey.

The ray traced pattern for Model 2 is shown in Figure 10a. The imaged zone extends from about 7000 to about 12 000 ft (2134–3658 m) TVD and is recorded by geophones that range from 8000 to 16 000 ft (2438–4877 m) TVD. This depth of imaging complements the far offset shot in Model 1 so that we anticipated good delineation of the salt flank geometry between the two surveys.

Model 2 was also used to assess the impact on imaging of an anhydrite-bearing cap on the top of the salt, if one were present. Our concern was that the cap would distort the rays and appreciably change the zone of imaging because anhydrite forms a very high velocity, highly refracting layer. Model 2 was altered slightly by introducing a cap with a maximum thickness of 400–500 ft (122–152 m) and a velocity of 18 kft/s (5486 m/s). The critical angle of refraction is 24 degrees at the top of the anhydrite layer. The ray paths under this variation of Model 2 are seen

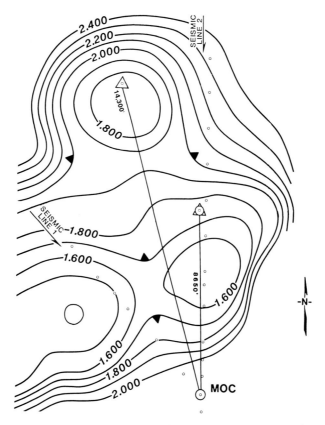

Fig. 8. Top of salt with location of Model 2 cross-section.

in Figure 10b. The rays traveling to the upper part of the borehole undergo the most refraction through the caprock, but the displacement of the ray paths is very small because the caprock is thin. The rays traveling to the lower part of the borehole are not affected because they enter the caprock near normal incidence. The zone of imaging is changed little by the cap and we did not anticipate problems because of the cap. No evidence of a cap was seen in the seismic data so an assumption was made that a thick cap was not present.

There were other models created that are not presented in this paper. These additional models tested the impact of changing the velocity model, sediment geometry, or salt top geometry. At the end of the exercises we felt very confident that most acquisition and interpretation difficulties had been anticipated. More importantly, we were confident that we could and would make correct interpretations of the field data when available.

SALT PROXIMITY RESULTS

A salt proximity survey was conducted according to the specifications of acquisition described in the previous section. The two survey shot locations are shown in Figure 8. A far-offset source, Station 1, was placed 14 300 ft (4359 m) from the well head. Station 2 was located 8650 ft (2637 m) from the wellhead. The entry point at the top of the salt, due to refraction in overlying sediments, would most likely occur on the high point of the salt crest seen along Line 2 and this high point is almost a horizontal surface.

A playout of the recorded time-depth panel is shown in Figure 11b from where first arrival times were picked. Figure 11a shows the output of our aplanatic program for the far offset shot. Seismic shot points and a time scale are posted on the display to facilitate comparison with seismic sections. The results are superimposed on a display of the model on which the analysis was performed. The Station 1 survey was acquired with geophone levels from 16 000 ft (4877 m/s) to

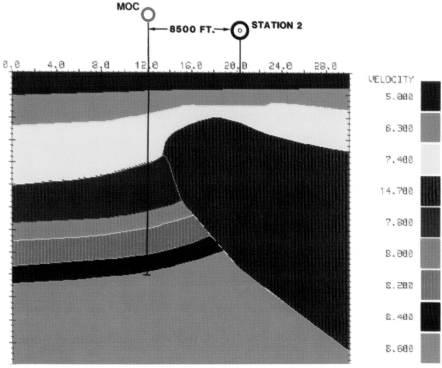

Fig. 9. Model 2 with shot station 8500 ft from borehole.

18 000 ft (5486 m/s) (the straight hole total depth was 18 500 ft) and sampled every
150 ft (46 m). This sampling interval honors the Nyquist requirement for the
source frequency and sediment velocities present. The interpreted curve indicates
that an overhang was detected by the survey.

The shallower survey, at Station 2, gave more details about the form of the
overhang. The aplanatic analysis is shown in Figure 12a and is based on first

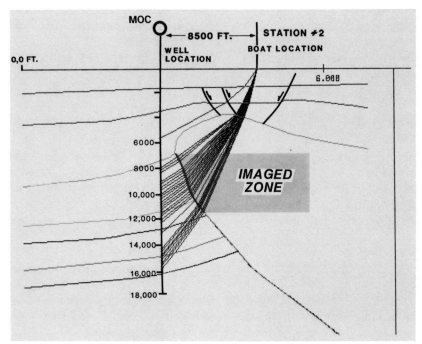

Fig. 10a. Raypaths for model 2 with shot offset of 8500 ft. Salt face is imaged between 7000
to 12 000 ft.

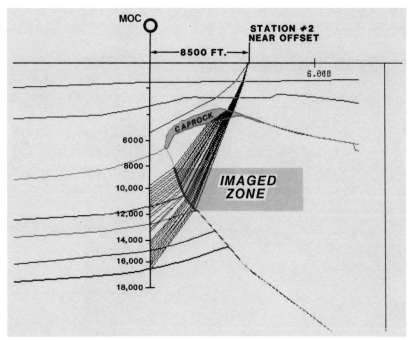

Fig. 10b. Ray paths for Model 2 modified with an anhydrite caprock. Salt face is imaged
between 8500 to 12 000 ft.

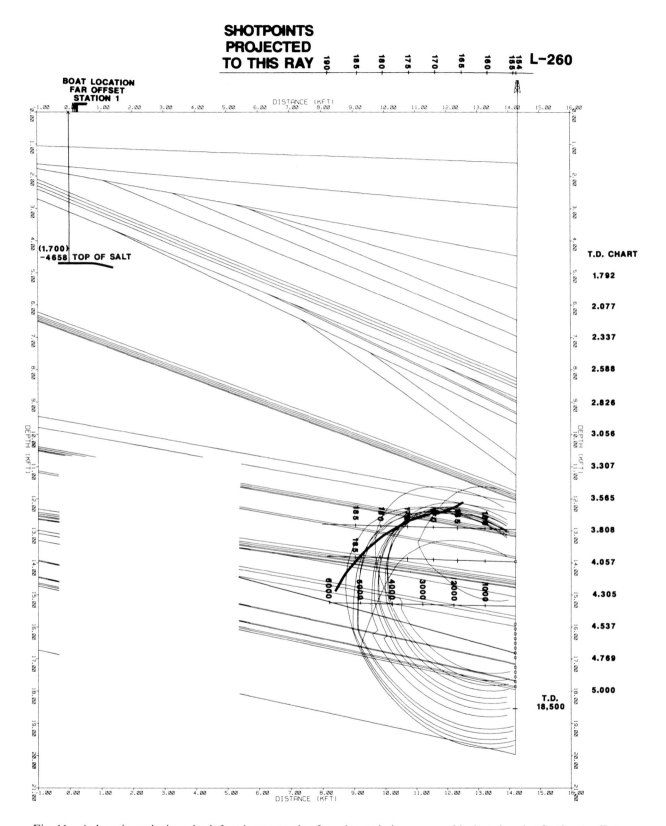

Fig. 11a. Aplanatic analysis and salt face interpretation for salt proximity survey with shot placed at Station 1, offset 14 300 ft from well.

arrivals taken from the time-depth panel of Figure 12b. The interpretation of the projecting nose and overhang of the salt are clearly displayed by the envelope of tangents to the aplanatic curves.

The combined salt face interpretation was converted to time and transferred to the seismic section Line 2, shown in Figure 13. The seismic interpretation of salt face location is shaded while the aplanatic results are posted as the two, thick, line segments at the edge of the shaded area.

A side track was drilled on the basis of this confirming evidence. The borehole trajectory of the side track is also drawn on Figure 13. The borehole penetrated salt at 3.3 s and exited salt at 3.6 s. The side track well encountered the overhang and thus confirmed the geophysical evidence.

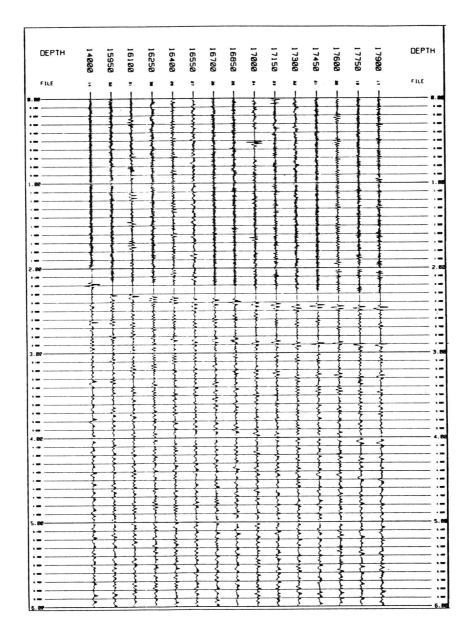

Fig. 11b. Record of vertical component geophones for Station 1.

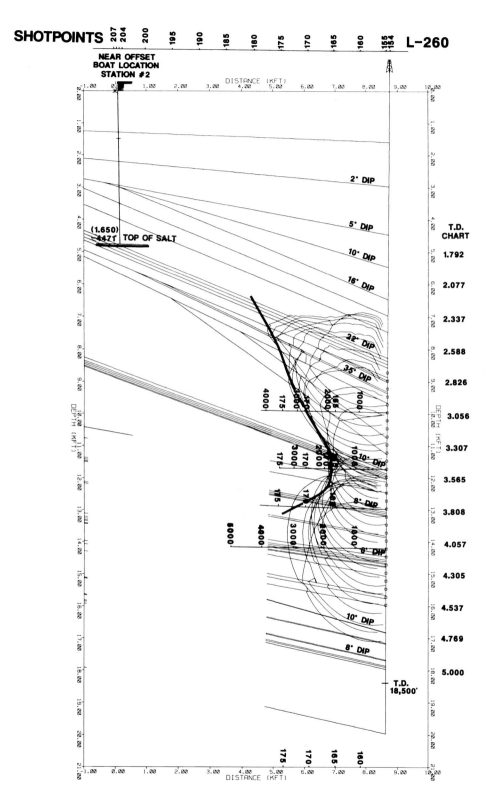

Fig. 12a. Aplanatic analysis and salt face interpretation for salt proximity survey with shot placed at Station 2, offset 8500 ft from well.

SUMMARY

Marathon Oil Company has routinely used the salt proximity survey to design side track boreholes on the flanks of salt domes. This paper shows one example of how modeling and ray tracing were used in support of the survey. Two computer models were created on the basis of the best available information before the survey was designed. Potential shot positions were evaluated for their efficacy in imaging the salt dome flank. In making these evaluations both potential source positions or some important model parameters were changed in order to observe the change in imaging that results.

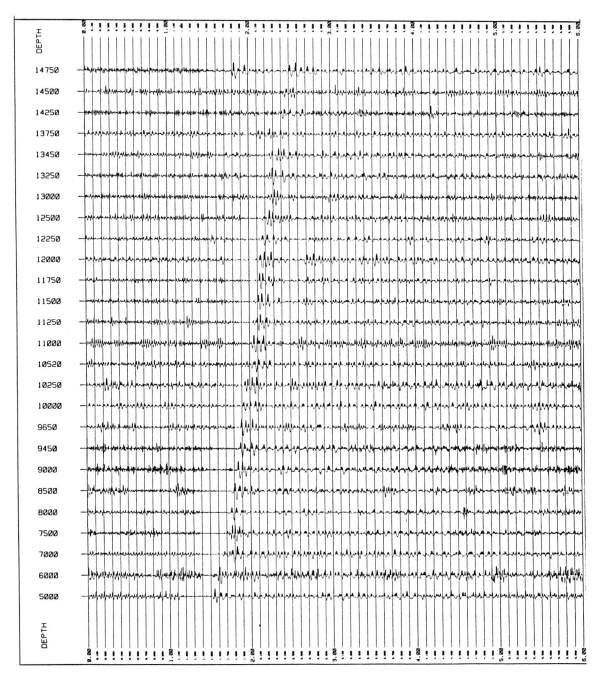

Fig. 12b. Record of vertical component geophones from Station 2.

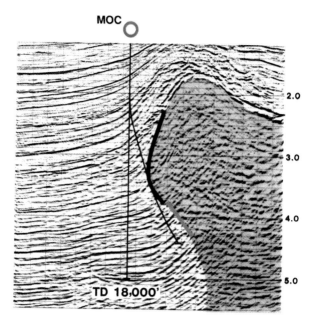

Fig. 13. Line 2 seismic section with well and side track trajectories shown. Shaded region is seismic interpretation of salt face. Thick lines are salt face interpretation based on salt proximity survey. Black and red lines are results from the near and far offsets, respectively.

Two shot locations were chosen for the acquisition program. The data and the aplanatic interpretations are presented, and the interpretation is posted on a seismic section. The sidetrack well is also displayed on the seismic section. The salt flank structure, defined by the proximity survey, was confirmed by the well. The salt proximity survey is a venerable idea that has been revived with the advent of 3-D seismic surveys and renewed prospecting around Gulf Coast salt domes.

ACKNOWLEDGMENT

The authors thank Marathon Oil Company for giving permission to publish this study. We also express our gratitude to Jerry Lowery for his invaluable support in the care and feeding of these computer programs. Our appreciation also is expressed to Ms. Julie Brock for helping us with the figures and Mr. Gary Ollenburger for the top of salt map.

REFERENCES

Gardner, L. W., 1949, Seismograph determination of salt-dome boundary using well detector deep on dome flank: Geophysics, 14, 29–38.
Musgrave, A. W., Wooley, W. C., and Gray, H., 1960, Outlining of salt masses by refraction methods: Geophysics, 25, 141–167.
Slotnick, M. M., 1959, Seismic interpretation theory, in Geyer, R. A., Ed., Lessons in seismic computing, Soc. Expl. Geophys., 3–9.

CASE HISTORY 6

Effective Depth Conversion: A North Sea Case Study

D. M. Zijlstra*, P. M. van der Made*, F. Bussemaker*,
P. van Riel*

INTRODUCTION

An important step in seismic interpretation is the time to depth conversion of an interpreted section. This conversion must be executed as accurately as possible, as errors may result in mispositioning of wells, incorrect estimation of closures, and miscalculation of potential reserves. A common approach to depth conversion, e.g., Marsden (1989), consists of two steps: (1) deriving velocities or velocity functions for the different layers; and (2) layer cake modeling of the time horizons, i.e., the procedure in which the layer velocity functions are used to convert the horizons to depth.

We recognize two potential sources of errors in this procedure. First, deriving the velocity functions is difficult. Well information is usually restricted to structural highs and thus does not indicate how velocities behave in structural lows. Averaging well information does not allow detection of velocity variations in the regions between the wells. Methods, which use stacking velocities to get more velocity information between wells, break down in areas of structural complexity.

The second source of errors lies in the fact that the velocity estimation and depth conversion are separately executed: the velocity functions depend on the depths of the horizons, however, the time-to-depth conversion is done after the velocity estimation. These difficulties could be avoided by using a one step procedure based on layer stripping, in which the depth conversion and the estimation of the velocity field, are executed simultaneously.

In this method each layer is inversely raytraced successively from the top down, and predicted moveout values are compared with observed ones. Unfortunately, the method has several serious drawbacks:

progressive error accumulation,

possible instability,

inability to properly handle features such as faults and unconformities, and

considerable user interaction is required.

*Jason Geosystems b.v., Delft, Netherlands.

Finally, there is a shortcoming common to all available methods: lack of quality control. Whether the output is consistent with the input data can not be verified, nor can the reliability of the depth conversion be determined.

The technique presented overcomes these problems, and is applied to an area in the North Sea to define structure below a salt ridge. In this technique, based on inverse raytracing, the derivation of layer velocities and depths is carried out in a single, automatic step. The boundaries of the velocity layers are depth converted horizons and other geologic features, such as faults and unconformities.

THE METHOD

In Figure 1a the logical flow of the inverse raytracing method is shown. The input data are

time picks of the interpreted horizons and the uncertainties in these picks;

stacking velocities at the interpreted horizons and the uncertainties in these velocities;

interpretive information on faulting and horizon termination describing how horizons and faults are connected to define a geologically meaningful structure.

The term "interpretation" specifically refers to this interpretive information as well as to the unmigrated horizon time picks.

"Velocity field" refers to a set of layers characterized by interval velocities, e.g., Figures 4 and 6. The boundaries of the layers are in depth and correspond to the depth-converted (migrated) time horizons and any structural features of the interpretation.

Fully Automatic Depth Conversion (FADC)

FADC, from the user's perspective, is one single step in the system flow. In fact, FADC consists of several, automatically executed steps as shown in Figure 1b. The procedure uses the method developed in van der Made (1988).

In choosing a method to derive a starting velocity field for the inverse raytracing, one condition must be met: the starting velocity field has to be sufficiently close to the true field to ensure convergence of the inverse raytracing. In practice, this condition presents little difficulty. Various alternative methods may be used to derive the starting velocity field. We use Dix-derived interval velocities as they can be easily obtained and incorporated in the automated procedure and are usually close enough to ensure convergence. The next step, the automatic updating procedure, consists of:

Forward modeling; the calculation of reflection times and associated stacking velocities for comparison with the input data. The stacking velocities are calculated by fitting a hyperbola to a set of traveltimes generated by offset raytracing through the velocity field.

Comparison of the modeled data with the input data to assess the data mismatch.

If the data mismatch exceeds the uncertainties of the input data, the interval velocities and the geometry of the boundaries defining the velocity field are updated to improve the match. This update is made in such a way as to ensure that the new velocity field is geologically meaningful. The benefit of offset raytracing should be noted. Incorporating offsets implies that stacking velocities are properly modeled which is particularly important in the case of dipping reflectors.

When a velocity field is generated that is consistent with the input data to within the data uncertainties, or when there is no longer convergence the automatic procedure is terminated.

Output

The output results are

a velocity field. (The boundaries of the velocity units of the velocity field correspond to the depth-converted and migrated interpreted events. Surfaces are truncated against faults and unconformities, in a manner consistent with the original interpretation).

quality control information on the match of the modeled and input traveltimes and stacking velocities in relation to the input data uncertainties.

quality control information on the reliability of the derived velocity field by means of a ray coverage plot.

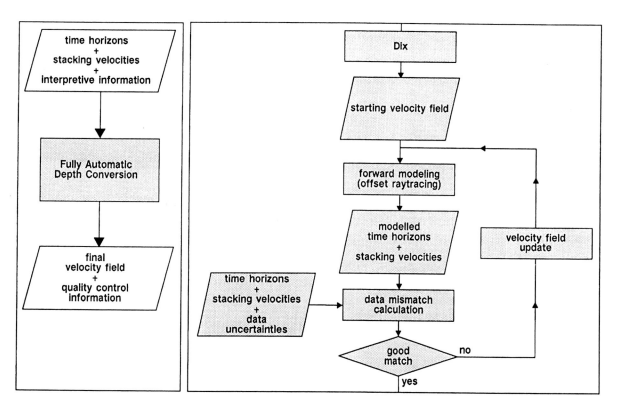

Fig. 1a. The general flow.   Fig. 1b. The main steps in the fully automatic depth conversion.

FIELD DATA EXAMPLE

Geology

In the North Sea several areas with hydrocarbon potential are characterized by the complex structure associated with mobilized salt. Figure 2 shows a typical example, from the platform along the western margin of the Central Graben. The main structural feature is a collapsed salt ridge overlying tilted block faults of Rotliegendes and an older section. Also, clearly visible at the right side of the section, is a Lower Cretaceous angular unconformity. The main exploration objective is the subsalt Rotliegendes section.

Input data

The seismic section in Figure 2 is unmigrated and shows the interpreted horizons. The picks of these horizons, which are used in the inverse raytracing, are shown in Figures 3a and 3b. Figure 3a shows the two-way traveltimes and Figure 3b shows the stacking velocities at the reflectors. The purple bars show the picking uncertainty of the data points. Note that, as indicated by their length, the uncertainties are allowed to vary with depth and per horizon in order to properly represent the degree to which the interpreter trusts the picks. In particular this is used to denote the uncertainty of the picks of the top Rotliegendes. Also note the red crosses in Figure 3a. These picks are tentative, and the interpreter has indicated that they are not to be used to drive the inverse raytracing. They are only used to complete the interpretation.

Depth conversion

The starting velocity field obtained by applying the Dix formula to the data in Figures 3a and 3b is shown in Figure 4.

To assess the quality of this depth conversion, offset raytracing is executed and the modeled traveltimes and stacking velocities are compared with the input data. This quality control result is shown in Figures 5a and 5b, where the modeled data values are shown as blue circles.

The result shows that, except for parts of the shallowest horizons, the blue circles lie outside the purple uncertainty bars in nearly every case. In other words, the starting interval velocity field is unsatisfactory and should be updated. To do this, the previously described, automatic inverse raytracing method is used.

Final result

The final depth conversion result obtained and the associated velocity field are shown in Figure 6. Note that all interval velocities are allowed to vary laterally. (The velocity values given are at the left and right ends of the section. The interval velocity at a certain position is found by interpolating the two values.) Significant changes are made to the interval velocities and the layer geometry of the initial velocity field. To judge this new result, three plots are generated for quality control.

The first two plots, Figures 7a and 7b, as already described, assess the ability of the final velocity field to recreate the observed zero offset traveltimes, and stacking velocity values. The third plot, Figure 8, shows the zero-offset ray coverage in the final depth conversion result. In this way the reliability of the final result can be assessed. In general, thick layers containing many and relatively long raypaths are well resolved for interval velocity and geometric structure. The reliability decreases for layers with few rays or for thin layers.

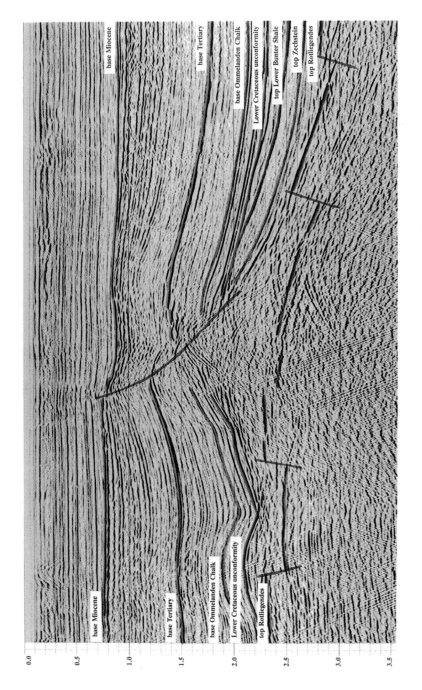

Fig. 2. Unmigrated seismic data and interpretation.

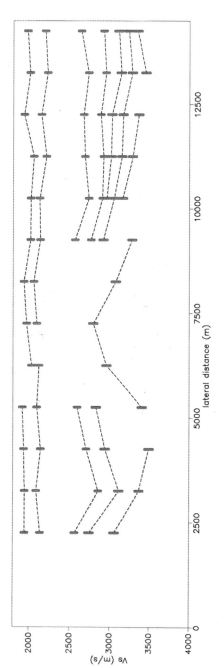

Fig. 3a. The input time picks and uncertainties (purple bars).

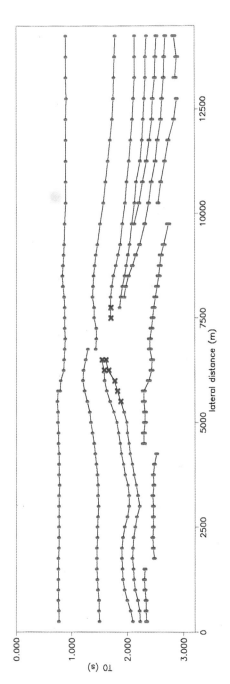

Fig. 3b. The input stacking velocities and uncertainties (purple bars).

1975.	m/s
2330.	m/s
2330.	m/s
4321.	m/s
4880.	m/s
5448.	m/s
5139.	m/s
5249.	m/s
4955.	m/s
5019.	m/s
4000.	m/s

Fig. 4. The initial velocity field. Note that the boundaries of the velocity units correspond to the geologic interpretation.

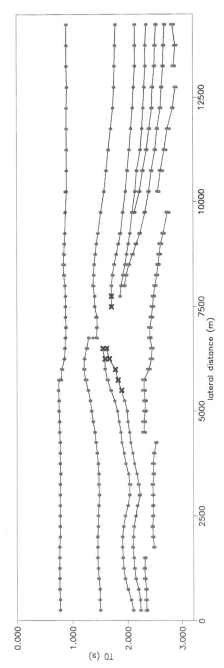

Fig. 5a. The input time picks and uncertainties (purple bars) and calculated time picks (blue circles) for the initial velocity field.

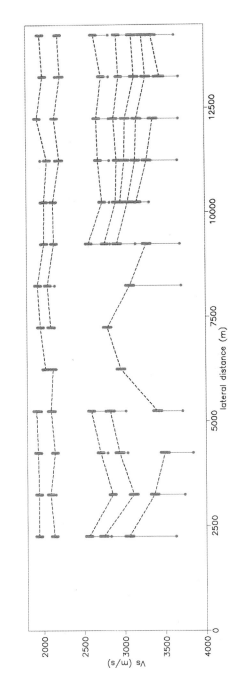

Fig. 5b. The input stacking velocities, uncertainties, and calculated stacking velocities (blue circles) for the initial velocity field.

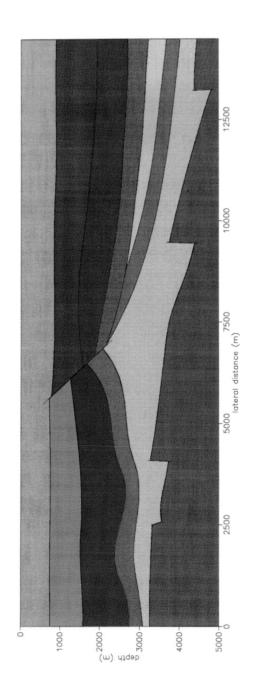

1923.	2017.	m/s
2327.	2048.	m/s
2216.	2384.	m/s
3746.	3810.	m/s
3310.	4494.	m/s
4918.	4430.	m/s
3341.	4944.	m/s
4246.	4890.	m/s
3725.	4326.	m/s
4324.	4803.	m/s
4000.	4000.	m/s

Fig. 6. The final velocity field. Note, in particular, the repositioning of the major fault with respect to the starting velocity field in Figure 4.

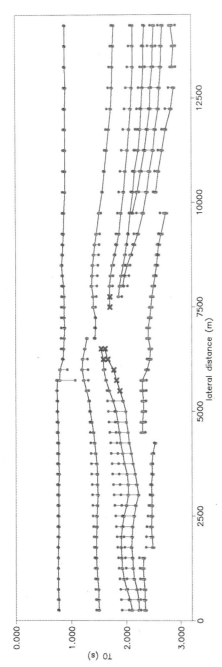

Fig. 7a. The input and calculated time picks for the final velocity field. Note that now the blue circles lie within the purple bars.

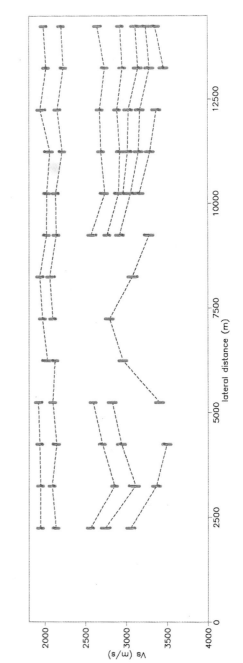

Fig. 7b. The input and calculated stacking velocities for the final velocity field. Note that now the blue circles lie within the purple bars.

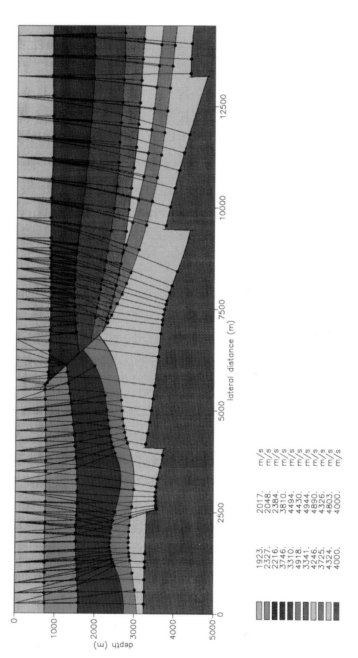

Fig. 8. The ray coverage plot of the final velocity field.

Figures 7a and 7b show that the mismatches are now within the specified data uncertainties almost everywhere, which implies that, based on the data match, there is no reason to reject the final result. When comparing the mismatch plots of the starting velocity field with the result of inverse raytracing (Figures 5a, 5b, 7a and 7b), the conclusion is certainly justified that the described inverse raytracing method is a powerful tool for time to depth conversion.

The third quality control plot, the zero offset ray coverage, is given in Figure 8. The series of layers terminating against the normal fault in the middle part of the section are shown to have a high coverage, which supports that part of the depth conversion result. The Lower Cretaceous unconformity on the left side of the fault has limited coverage near the fault. Therefore, only limited confidence can be put in the structural estimation of this unconformity near the fault.

Subsequent comparison with an off-line, but nearby well shows good results. All depths, except for the Rotliegendes are correct within a margin of 0.5–2 percent. The heavily faulted top Rotliegendes, the deepest reflector, shows a deviation of 3 percent which reflects the uncertainty of the picks as specified by the data uncertainties.

DISCUSSION

One important, advantageous feature of the method is not easily visualized: the benefit of indirect resolution. This benefit is due to the additional information supplied through the requirement of geologic consistency of the depth-converted interpretation produced by the method. In particular, this is the case with terminating horizons. When the horizons dip, the inverse raytracing procedure will migrate them toward their correct position. Because the requirement of geologic consistency must be honored, the fault or interface to which the terminating horizon is connected will be migrated in the correct direction.

In the case discussed, the indirect resolution feature is clearly seen at the main fault. Even though no reflection energy is present from the fault interface, the dip is greatly improved compared to the initial estimate shown in Figure 4. The presence of diffraction energy, which comes from the termination point of a reflector, is important in defining the location of the fault. An example is visible at the termination of the base Tertiary on the footwall side of the fault (see Figure 8) where the point clearly acts as a diffractor.

The example demonstrates that the inverse raytracing method described is a powerful tool in depth conversion:

> The complete interpretation is converted to depth in a single, automatic step, irrespective of the structural complexity.

> The correct velocity field does not need to be specified as input. A velocity field matching the time picks and stacking velocities is automatically generated. The horizons and structural features are defined by the boundaries of the velocity units and are consistent with the interpretation of truncation relationships associated with faults and unconformities.

> Easy to interpret quality control information is generated to assess the quality of the result.

Of course, a method can only be as good as the data used for input. The ultimate limitation of the method's applicability lies in the accuracy of stacking information. Use of this information for estimating complex velocity fields is not always looked upon favorably. However, to conclude that problems with velocity estimation methods which use stacking information are due to the stacking information is not correct. The problem is in the methods themselves, which do not properly account for how the stacking process works. There are several reasons why the described inverse raytracing method can be expected to perform well in areas where alternative techniques fail:

> Layer cake modeling is not used; ray path bending is properly simulated by raytracing.

> In several approaches, obtaining a velocity field and migrating time horizons to depth are considered separate problems. These problems are, of course, closely related. The proposed method honors and makes use of this relationship to obtain a more reliable result.

> The stacking velocities are directly calculated through offset raytracing. In some alternative methods, stacking velocities are approximated using the local wave front curvature around the zero-offset ray, and assume no reflection point smearing in a common midpoint (CMP) gather. With high velocity contrasts and/or steep dips this approximation breaks down much more readily than the stacking velocities estimates.

> The dip effect in stacking velocities is properly handled and actually used to obtain a better depth conversion result.

> The output is a geologically consistent interpretation. As was discussed, use of this additional information means that good quality data outside a zone of high structural complexity are indirectly used to resolve features inside such a zone.

> All interval velocity and geometry parameters are simultaneously estimated instead of recursively, as is the case in methods which make use of layer stripping. This estimating implies that information from deeper reflectors also is used to estimate the parameters of the shallower part of the velocity field. In comparison with layer stripping, this property results in an improvement of the overall accuracy of the final depth interpretation and is a natural safeguard against error accumulation with depth.

As a result of these characteristics the method is able to handle a wide range of time to depth conversion problems. We have applied the method to an extensive set of different cases and found that the depth converted results match the data from wells within a margin of 0.5–3 percent. In these cases we did not use well data as input; however, the method lends itself very well for the incorporation of this kind of information as well as geologic data such as known velocities, velocity gradients, and thickness.

CONCLUSION

In interpretation an important step is the time to depth conversion of the final interpretation. Though several methods for depth conversion are available, there are few which produce satisfactory results in areas of structural complexity.

Inverse raytracing, as applied in the method presented, can be successfully applied in many such cases. The complete interpretation is automatically converted to depth in one step, using an automatically generated velocity field which is consistent with the interpretation. The reliability of the result may be judged from easily inspected plots with quality control information.

ACKNOWLEDGMENT

Jason Geosystems thanks Unocal Netherlands b.v. for kindly providing the data.

REFERENCES

van der Made, P. M., 1988, Determination of macro subsurface models by generalised inversion: Ph.D. dissertation, Delft Univ.
Marsden, D., 1989, Layer cake depth conversion: The Leading Edge, **8**, 10–14.

CASE HISTORY 7

Modeling the Seismic Response of Geologic Structures with Physical Models

*Linda J. Zimmerman**

INTRODUCTION

For a given exploration problem, an explorationist must choose the most accurate and yet cost-effective method for interpreting the seismic response of an earth section. An accurate interpretation translates the measured round-trip traveltimes of seismic waves into a picture of the subsurface. Wave-equation theories link traveltimes with depth. However, interpreting seismic data in terms of wave propagation theories is difficult because rarely is a sufficiently detailed knowledge of all the relevant parameters available. For example, without accurate velocity information, a unique depth picture is impossible. Deriving accurate velocity information from seismic data can be very difficult. Therefore, interpreters frequently compare the seismic response of probable models of the subsurface which have known elastic parameters and structures with a field data set to test an interpretation. A match between model data and field data increases an interpreters confidence in the picture of the subsurface inferred from the seismic data.

The inherent ambiguity of seismic data is magnified in complexly structured areas. When a structure of interest is three-dimensional (3-D), very large, and very complex, a fully elastic, 3-D model is necessary. Both physical and computer modeling methods are valid tools for studying the propagation of elastic waves through 3-D, layered subsurfaces. However, the two types of tools are not equally suited for studying all types of problems. Ray-tracing computer programs are fast, accurate, and inexpensive tools for studying systems of medium-sized, smoothly varying structures when amplitude responses are not critical. More computer-intensive, fully elastic, wave-equation programs are needed if amplitudes are important. Unfortunately, wave-equation methods are currently limited to either small models or flat-layered subsurfaces. Even with modern supercomputer technology, the amount of computer time and memory required to collect high-fold, 3-D data sets on models of large complex structures is prohibitively expensive. Currently, only a solid physical model can accurately represent the seismic response of very large and very complex structures for a reasonable cost.

*Exxon Production Research Company.

PHYSICAL MODELS

Seismic interpreters have been using physical models to study wave propagation since the early part of the century (Terada and Tsuboi, 1927). Physical modeling flourished during the 1950s with researchers using both 2-D (Oliver et al., 1954) and 3-D (Levin and Hibbard, 1955) models. But even at that time, physical models were hampered by their inherent inflexibility as well as by a lack of suitable digital recorders. Consequently, by the early 1970s advances in computer technology were rapidly making physical models seem obsolete. The development of high-speed, high-resolution digital recorders in the early 1980s combined with a renewed interest in exploring for hydrocarbons in areas with very complex structure resurrected an interest in physical seismic modeling.

Physical models are especially valuable for collecting large data sets with unusual data acquisition configurations. Although the shape and physical parameters of physical models are difficult to change, data acquisition parameters can be varied easily and inexpensively. Physical models behave enough like the earth that important features in field surveys can be reproduced; yet a physical model can be simplified without giving up too much realism. Therefore, to separate effects and study them individually is possible, although the separation is usually impossible with field data. This flexibility permits extensive data acquisition tests before an expensive field study is initiated.

Physical models are scaled to keep the size of the model manageable while maintaining the desired resolution. Normally, scale factors are chosen to accommodate the bandwidth of the model sensors, the digitizer speed, and the limitations on constructing thin beds. Typical scale factors range between 1000:1 and 10 000:1. Usually, distance is scaled down and frequency is scaled up while velocity is left unscaled. However, to model very large land areas or very high-velocity sections, velocity scaling is also an option. One difficulty with scale models is the relatively large size of the sources and detectors after scaling. However, experiments have shown that some relatively large, areal sources act approximately like point sources when compared with the mathematically calculated response (Gray, 1986).

Either land or marine data can be collected on a physical model using piezoelectric crystals as the sources and detectors of acoustic waves. These devices are usually operated between 10 and 500 KHz. Marine detectors are coupled to a model through a liquid surface layer. Land sensors are coupled directly to the surface of a model with pressure. Collecting land data on physical models is complicated by high amplitude surface waves which tend to obscure the reflected energy. Since the resolution of high-speed, digital recorders is limited, surface waves must be mechanically damped before recording. One method of damping surface waves is by adding a thin, viscous, surface layer to the model (Gray, 1986).

PHYSICAL MODELING FACILITIES

Some universities, including the University of Calgary and Delft University of Technology, have physical modeling facilities. The University of Houston's Seismic Acoustic Laboratory has an active physical seismic modeling research program. In the past, the majority of their physical seismic modeling was done with a water tank modeling system where only P-wave data are collected (French, 1974). However, recently they have published papers describing work with solid models including models to study the effects of surface waves, shear waves, and anisotropy effects (Tatham et al., 1987).

The largest physical modeling facility currently in use is probably the one at

Exxon Production Research Company in Houston. Exxon's facility can be used to model nearly any earth section of exploration interest. Models of earth sections as large as 100 square miles to depths of 20 000 ft, with complexly structured beds as thin as 1/10 wavelength, have been constructed. The models are built from very fine-grained cement mixtures and polyester resins. These construction materials provide a nearly continuous range of unscaled velocities between 4000 and 12 000 ft/s, with densities between 0.8 and 2.2 g/cc (Gray, 1986).

EXAMPLE OF A PHYSICAL MODEL

A photograph of one of Exxon's physical models is shown in Figure 1. The subsurface reflectors are marked on the side of the model. The model measures 10 ft by 12 ft by 3 ft thick, which represents a land area of about 100 square miles to a depth of 20 000 ft. The distance scale factor for this model is 6600 to 1. The elastic parameters of the model layers were chosen to simulate velocity profiles and impedance contrasts observed in wells drilled near the area being modeled. Figure 2 is a cross-sectional sketch of the model showing the scaled P-wave velocity of each layer. The model was used to test novel data acquisition configurations that would have been prohibitively expensive to test in the field.

The model was based on an interpretation of a 3-D seismic survey over a complexly deformed, Gulf of Mexico salt sill. The structure of the model was designed from contour maps drawn on major reflectors identified from the field survey. The contour design maps honored the field seismic data where reflectors were mappable. Where no reflectors were visible on the field lines, reflectors beneath the salt were assumed to be nearly conformable to the base of the salt. Figure 3 is a contour map of the top of the salt. Figure 4 is a photograph of the top of the salt as reproduced by the model.

Figure 5 shows a 48-fold, migrated line from the field survey. The location of the line is shown in Figure 3. The field data illustrate the sort of data quality problems typical of areas around salt sills in the Gulf of Mexico. Reflectors above and beside the salt are continuous and easily mappable. Reflectors below the salt are well imaged away from the salt structure, but rapidly lose signal where they continue below the sill.

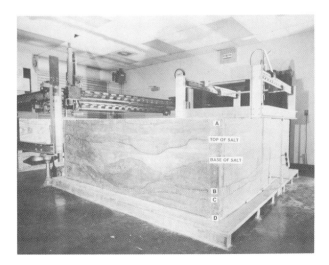

Fig. 1. A photograph of the physical model with one reflector above the top of the salt (A), the top and base of the salt, two reflectors below the level of the salt (B and C), and the base of the model reflection (D) labeled.

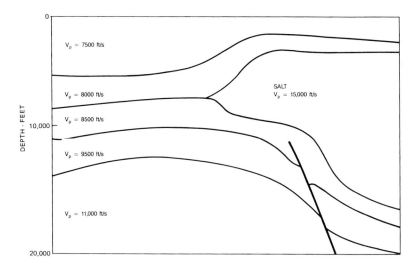

Fig. 2. Cross-sectional sketch of the physical model showing the *P*-wave velocity of each layer.

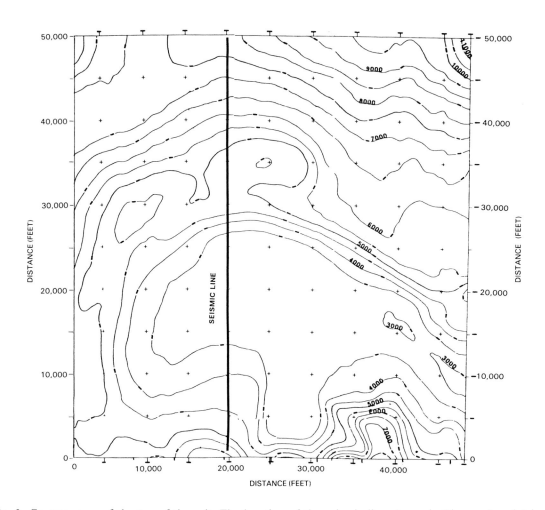

Fig. 3. Contour map of the top of the salt. The location of the seismic line shown in Figures 5 and 6 is shown superimposed on the top of the salt contours.

Figure 6 is a model line shot with identical acquisition parameters and at the same location over the salt as the field line shown in Figure 5. Ray trace computer models and two-dimensional acoustic wave equation models do not fully explain the loss of signal observed below salt in some locations. However, features in the physical model section are seen as similar to those observed in the field section. A conventional stack of the data images a reflector above the top of the salt (A), the top and base of the salt, two reflectors below the level of the salt (B and C),

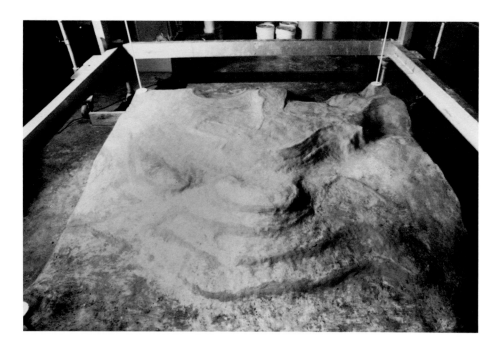

Fig. 4. Photograph of the top of the salt as it was reproduced by the model.

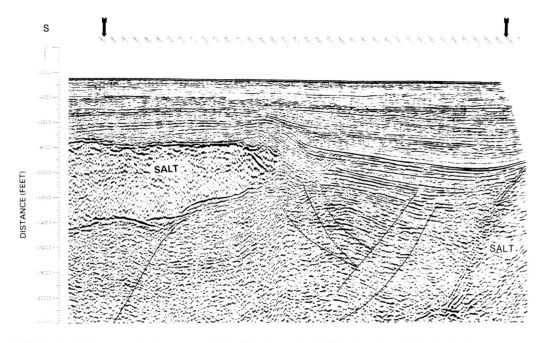

Fig. 5. Field seismic line acquired across a Gulf of Mexico salt sill. The location of the line is shown superimposed on the top of the salt contours in Figure 2.

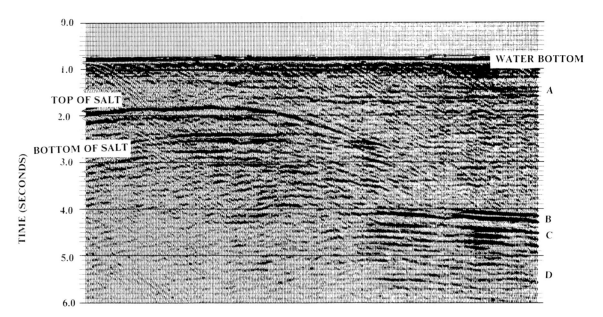

Fig. 6. Model seismic line shot with the same acquisition parameters and at the same location over the salt as the field line shown in Figure 5. One reflector above the top of the salt (A), the top and base of the salt, two reflectors below the level of the salt (B and C), and the base of the model reflection (D) are labeled. Line location is shown superimposed on the top of the salt contours in Figure 2.

and the base of the model reflection (D) beside the salt. As in the field data, reflectors visible beside the salt vanish beneath the salt.

An analysis of these data demonstrated a strong similarity in signal characteristics between physical model data and field data when the field data acquisition parameters were duplicated. The similarity between the two data sets implies use of the model for testing novel data acquisition schemes to enhance subsalt reflectors is valid. Furthermore, the model can be used to investigate the reasons why the reflection signal diminishes so greatly below salt in some areas and not in others. Success in enhancing subsalt reflections on the model data increases the probability a technique will succeed in the field.

CONCLUSION

Good agreement between physical model data and computer-generated data for a simple system (French, 1974; Gray, 1986) and the apparent similarity between field data and physical model data imply that use of physical models to model earth sections too complex to be adequately modeled with computer methods is reasonable. However, we can anticipate that the increasing power of digital computers and the increasingly sophisticated software for simulating wave propagation through complex subsurfaces will eventually make physical models obsolete. But, until that time, physical models will continue to serve as valuable aids to interpreters and researchers studying wave propagation through very large and very complexly structured subsurfaces.

REFERENCES

Gray, L. J., 1986, A comparison of three-dimensional seismic models: Expanded Abstracts, 2nd Ann. Gulf Coast Expl. and Develop. Mtg., Soc. Expl. Geophys.
French, W. S., 1974, Two-dimensional and three-dimensional migration of model experiment reflection profiles: Geophysics, 39, 265–277.

Oliver, J., Press, E., and Ewing, M., 1954, Two-dimensional model seismology: Geophysics, **19**, 202–219.

Levin, F. K., and Hibbard, H. C., 1955, Three-dimensional seismic modeling studies: Geophysics, **20**, 19–32.

Tatham, R. H., Matthews, M. D., Sekharan, K. K., Wade, C. J., and Liro, L. M., 1987, A physical model study of shear wave splitting and fracture intensity: 57th Ann. Internat. Mtg., Soc. Expl. Geophys., Expanded Abstracts.

Terada, T., and Tsuboi, C., 1927, Experimental studies on elastic waves (Parts I and II): Tokoyo Imperial University, Bull. Earthquake Res. Inst., **3**, 55–65, **4**, 9–20.

CASE HISTORY 8

Seismic Modeling of a Pinnacle Reef: An Example from the Williston Basin

K. W. Rudolph* and S. M. Greenlee*

ABSTRACT

By using two- and three-dimensional ray-trace seismic modeling, the subtle seismic response of a small Winnepegosis (Devonian) pinnacle reef from the Williston Basin was successfully modeled. Because of the small size and steep stratigraphic dips, the crest of the reef acts as a scattering point and generates a low amplitude diffraction on the unmigrated seismic data. Other seismic manifestations include disruption of deeper reflections, paired diffractions sourced from the toe of the slope, velocity pullup of deeper reflections, and nucleation on a structural terrace. Modeled seismic criteria are only interpretable on the actual data when the data is redisplayed using a squeezed gray-scale display. Synthetic seismic sections shot at the toe of slope and off-buildup positions show that recognition of the buildup crest is difficult without a line shot perpendicular to the axis of the hyperbolaes that represent the apparent reef crest. Three-dimensional ray-trace modeling also shows that pinnacles are detected by seismic sections shot off of the buildup. In these cases, two-dimensional migration may then result in the false imaging of a buildup along a line where none occurs. These modeling criteria are particularly useful in recognizing the occurrence and in mapping the crest of small carbonate buildups.

INTRODUCTION

Seismic modeling of stratigraphy generally uses techniques of amplitude modeling or seismic impedance to predict lithology or porosity. However, in cases where depositional dips are steep, the issue of raypath effects becomes critical. Carbonate buildups represent a common example of steep stratigraphic dips. Seismic recognition and prediction of the geometry of steep-sided buildups is a structural modeling problem, rather than a stratigraphic modeling problem.

The seismic line used in this study intersects a small pinnacle reef (Winnepegosis Fm.) that has been confirmed by drilling. This seismic line was studied using forward seismic modeling techniques to:

> Explain the subtle seismic expression of the buildup on the unmigrated seismic data.

*Exxon Production Research Company, P. O. Box 2189, Houston, Texas 77252-2189.

Develop seismic criteria for the recognition of small buildups on unmigrated seismic data.

Evaluate the ability of migration to image the buildup.

Investigate the effects of sideswipe on the recognition and imaging of small pinnacles in the subsurface by using three-dimensional (3-D) forward modeling.

STRATIGRAPHY AND HYDROCARBON PRODUCTION

The study area is in the Williston Basin, in Renville County, North Dakota. During the Middle Devonian, the northern portion of the Williston Basin was part of the greater Elk Point Basin, which extended into Saskatchewan and Manitoba, Canada. During deposition of the Winnepegosis Fm., the Elk Point was a moderately deep basin fringed by a carbonate shelf (Figure 1). Scattered on the basin floor are isolated Winnepegosis pinnacle reefs and small platforms.

The basal portion of the Middle Devonian is the Ashern Fm., a red bed succession that rests unconformably on the Silurian Interlake Fm. (Figure 2). Carbonates of the overlying Winnepegosis are interpreted to have been deposited during the succeeding relative rise of sea level. Shelf to basin relief of several hundred feet developed during deposition of the Winnepegosis due to differential carbonate sedimentation. Basinward of the shelf margin, Winnepegosis deposition was terminated by a relative fall in sea level (lower Givetian sequence boundary). Evidence of this sea level fall includes vadose meteoric diagenesis in the upper portions of Winnepegosis buildups (Kendall, 1975; Perrin 1982). During the lower Givetian lowstand, the Elk Point Basin became a restricted evaporitic basin. Restriction resulted in the deposition of thick halite, with ancillary anhydrite, of the Prairie Evaporite. The Prairie Evaporite is about 500 ft (150 m) thick in the study area and fills in depositional relief at the top of Winnepegosis and onlaps the flanks of the basin. The Prairie Evaporite is capped by a thin red bed and thick carbonate section of the Dawson Bay Fm.

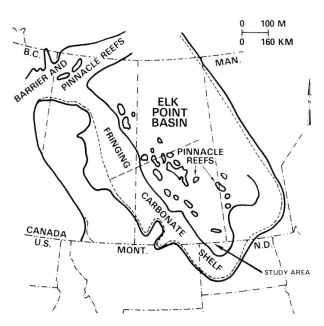

Fig. 1. Middle Devonian paleogeography, Elk Point Basin (after Grayston, et al., 1964).

Production from the Winnepegosis occurs in structural traps, where fracturing has enhanced permeability in low porosity dolomite (e.g., Outlook, Fairview, Reserve, Raymond, and Medicine Lake Fields). Production also has been established in isolated pinnacle reefs where closure is provided by depositional relief on the buildup (e.g., Norma and Tableland Fields). Two parameters that may prove critical in the pinnacle reef play are top seal and salt plugging. Winnepegosis pinnacles that developed maximum relief may extend above the Prairie Evaporite and lie in contact with porous Dawson Bay carbonate. Another risk is salt plugging of the porosity, with halite being introduced from the Prairie Evaporite.

The primary objective of this study is to model the seismic response of a small Winnepegosis pinnacle reef. The seismic line used in this study intersects two wells that penetrate the Winnepegosis (Figure 3). The Shell, Golden 34x-34 drilled an interpreted Winnepegosis pinnacle reef and is completed as an oil well in the upper part of the Winnepegosis. The Shell Golden 44x-34 was drilled 850 ft to the east and encountered a normal off-reef section of Winnepegosis. This well was plugged and abandoned. Based on Winnepegosis thickness changes, there are approximately 220 ft of depositional relief between the two wells at the top of the Winnepegosis.

STRATIGRAPHIC TO SEISMIC DATA CALIBRATION

The velocity structure of the Middle Devonian in the study area is illustrated in Figure 3. The Winnepegosis section has some variability due to porosity devel-

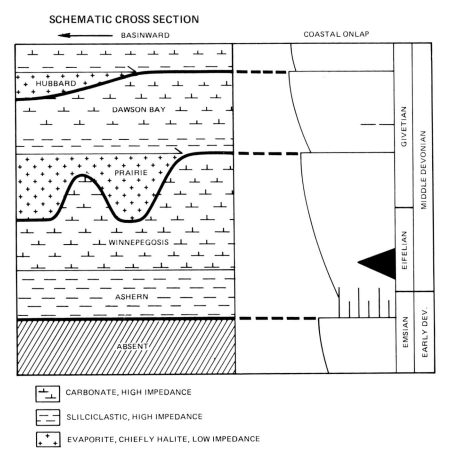

Fig. 2. Middle Devonian sequence stratigraphy.

opment, but is generally high impedance (V = 17 000 ft/s, d = 2.7). The Prairie Evaporite is low impedance (V = 14 000 ft/s, d = 2.05) and the Dawson Bay is high impedance (V = 18 500 ft/s, d = 2.7).

The seismic data employed in this study contain a zero-phase pulse with a central peak (black) corresponding to a positive reflection coefficient. Therefore, the surface at the top of the Prairie Evaporite corresponds to a negative reflection coefficient and the center of a trough (white) on the seismic data (Figures 4 and 5). The top of the Winnepegosis corresponds to a positive reflection coefficient and the center of a peak (Figures 4 and 5).

FORWARD SEISMIC MODELING

Examination of the unmigrated seismic line that ties the two wells gives little indication of a seismic anomaly associated with the buildup (Figure 4). There is a slight apparent sag at the reflection that corresponds to the top of Winnepegosis in the off-reef well (44x-34).

A key consideration is that these data are not migrated. Most data from the Williston Basin are not migrated because structural dips are generally very low. However, in this example, there is strong evidence of steep *stratigraphic* dips from the well data (Figure 3): 220 ft (66 m) of relief over a maximum distance of 850 ft (255 m) at the top of Winnepegosis.

A 2-D depth model constructed from this seismic and well information shows the raypath effects of a small pinnacle with steep flanks (Figure 6). The crest of the pinnacle and perhaps the toe-of-slope acts as a scattering or diffraction point because of their small radius of curvature. In addition, the model shows that reflections from the reef flanks may be difficult to record because of their steep dip.

The synthetic seismic section created by this ray-tracing is shown in Figure 7.

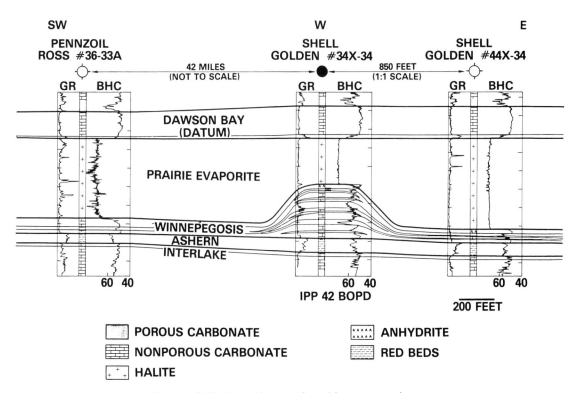

Fig. 3. Middle Devonian stratigraphic cross section.

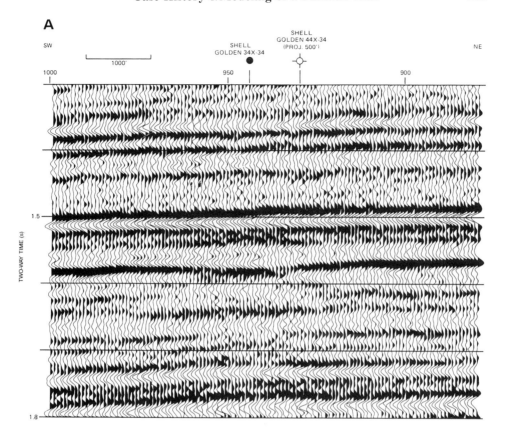

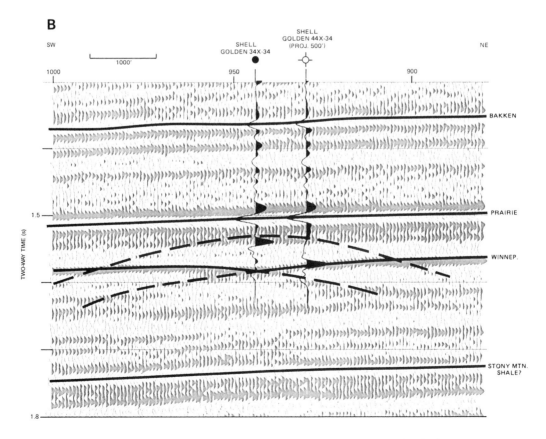

Fig. 4. Uninterpreted (A) and interpreted (B) unmigrated seismic line.

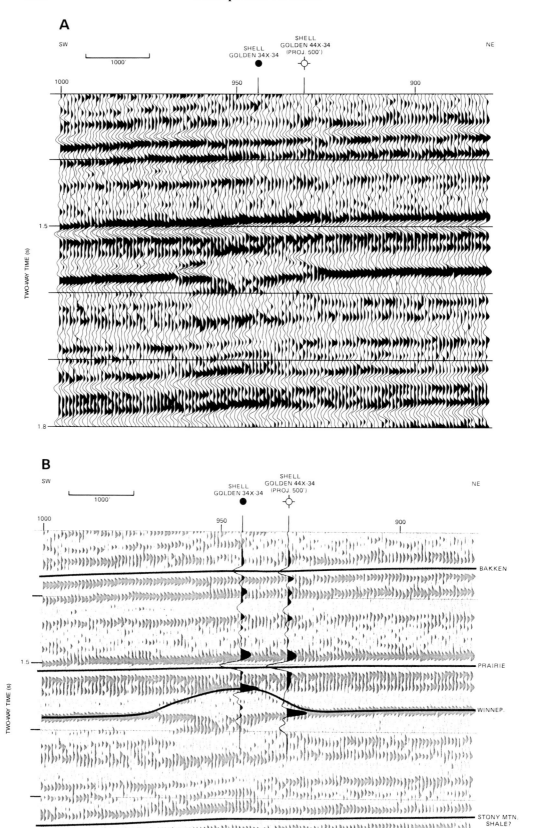

Fig. 5. Uninterpreted (A) and interpreted (B) migrated seismic line.

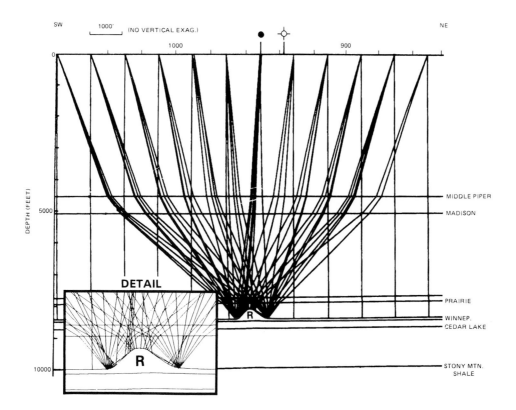

Fig. 6. Normal incidence raypaths from top of Winnepegosis.

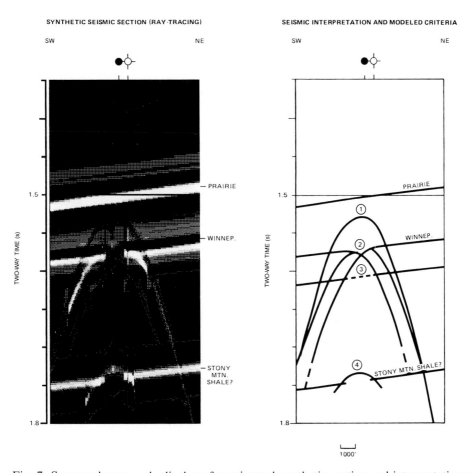

Fig. 7. Squeezed gray scale display of unmigrated synthetic section and interpretation of events.

It is most effectively compared with the actual data (Figure 8) on squeezed gray scale displays. Several seismic features can be recognized and related to the buildup geometry:

Diffraction at reef crest—scattering due to steep curvature.

Paired diffractions at toe-of-slope—scattering at abrupt change in dip.

Disruption of deeper reflections—interference with diffractions and raypath distortion through reef.

Nucleation of reef on structural terrace—subtle change in structural dips of deep reflections occurs at the position of reef. This change may indicate that the buildup was nucleated on an antecedent structural anomaly.

All of these features are represented in the real data. However, the base of the buildup was not well simulated. On the synthetic data the buildup acts as a lens that severely deteriorates the amplitude of the reflection at the base. On the real data this reflection is of similar amplitude to the reflection off the buildup. Interestingly, both the synthetic and real data depict a slight pull-up below the buildup along the deep Stony Mountain Shale reflection.

MIGRATION

Ray-trace modeling documents the seismic manifestation of the Winnepegosis pinnacle reef on the unmigrated seismic data. An alternative is to migrate the data,

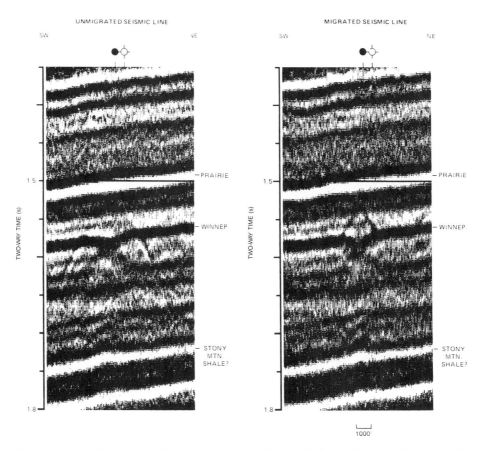

Fig. 8. Squeezed gray scale display of unmigrated seismic line and migrated seismic line.

which should collapse diffractions to their scattering points and move dipping reflections to their proper position. The seismic line was migrated using horizon-keyed interval velocities derived by merging interpreted seismic sequences and well velocities. The migration succeeded in collapsing the diffractions and imaging a buildup at the position of the reef (Figures 5 and 8).

However, the migration has not perfectly imaged the buildup. The migrated seismic data indicates that the 44x-34 well is on the lower flank of the buildup instead of its actual off-reef position. Also, the seismically defined toe-of-slope does not correspond to the position at which the top of Winnepegosis reflection in the off-reef becomes low amplitude. The failure of the migration to perfectly image the reef may be due to one or more factors: limitations of the migration algorithm, steep flank dips inadequately recorded during acquisition, or 3-D effects.

THREE-DIMENSIONAL FORWARD MODELING

Wells drilled on well-defined but small buildup anomalies frequently come in low-to-prediction or on regional dip. This result is caused in many cases, we believe, by 2-D migration of buildup sideswipe reflections. 3-D model studies of a reef are discussed in Neidell and Poggiagliolmi (1977). They note the Fresnel-zone effects of on- and off-buildup seismic lines which we have not modeled here. A 3-D model of a 500 ft (95 m) high buildup, 4000 ft (1220 m) in diameter and buried to 7000 ft (2134 m), is shown in Figure 9. The paired diffractions generated at the toe-of-slope in the 2-D Winnepegosis model are not included in the 3-D model

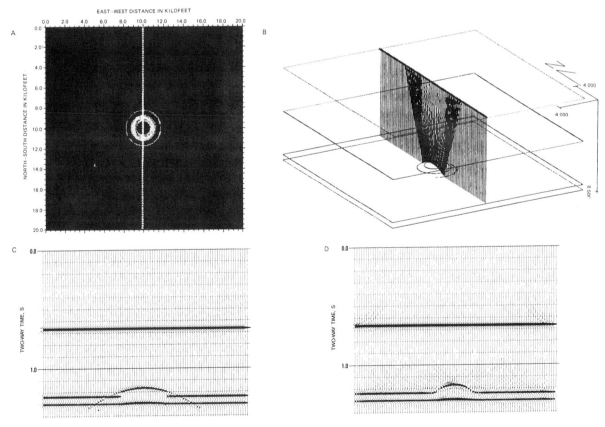

Fig. 9. Contour map with ray paths superimposed (A), three dimensional perspective plot (B), unmigrated synthetic seismic section (C), and migrated synthetic seismic section (D) for line shot over the crest of the buildup.

because of program limitations. Four synthetic seismic lines were shot over the model. These lines simulate a seismic line shot over the crest of the buildup, one in a mid-flank position, one near the toe-of-slope, and one 2000 ft (610 m) off the buildup. Flank reflections recognized from the crestal synthetic seismic section (Figure 9) can be migrated into their proper positions. Plots showing the normal-incidence rays captured by the mid-flank line (Figure 10) indicate that energy is gathered from near the crest of the buildup. Both the unmigrated and the migrated sections show arrivals for the buildup top significantly higher than the model positions in the line of section. In fact, a comparison of the reef crest and mid-flank migrated seismic lines shows them to be nearly identical. A similar phenomena is illustrated by the toe-of-slope seismic line (Figure 11). Although the line passes over only 50 ft (15 m) of buildup relief, it suggests over 350 ft (107 m) of buildup is present. Again the error is due to sideswipe. A seismic line shot 2000 ft (610 m) from the toe-of-slope still shows a mounded seismic anomaly due to sideswipe from the reef flank (Figure 12).

From this analysis, we conclude that additional data should be acquired perpendicular to the axis of the apparent crest of the buildup whenever possible. In the case of very small buildups, only 3-D seismic data calibrated with modeling studies may accurately delineate the buildup. Such was the case in the development of the Humble City South Field in Lea County, New Mexico (Caughey, 1988).

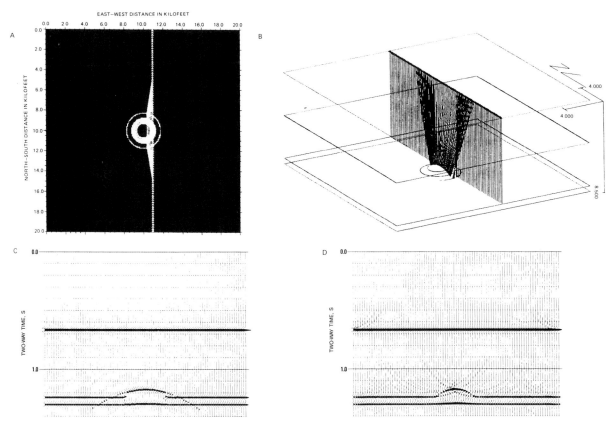

Fig. 10. Contour map with ray paths superimposed (A), 3-D perspective plot (B), unmigrated synthetic seismic section (C), and migrated synthetic seismic section (D) for line shot midway down the flank of the buildup model. Note that normal incidence rays captured by the seismic line originate near the crest of the buildup. Because of this, the resultant unmigrated and migrated seismic sections are almost identical to the synthetic seismic lines shot over the crest of the buildup.

CONCLUSIONS

Steep and complex stratigraphic geometries pose the same challenges that we encounter in seismic interpretation of complex structure:

Reflections out-of-place on unmigrated data (e.g., diffractions from reef crest and toe-of-slope).

Reflection energy captured from out of the plane of section (e.g., sideswipe of reef crest).

Raypath distortions due to complex geometry (e.g., disruption of events below reef).

Velocity effects due to rapid changes in velocity field (e.g., velocity pullup below reef).

Inadequate recording and imaging due to steep dips (e.g., flanks of reef).

Imperfect migration due to limitations of data or algorithm (e.g., imaging of reef flanks).

These issues have been addressed in this study by using ray trace forward modeling to model seismic arrival times and amplitudes from the unmigrated data. If the only stratigraphic issue is geometry, inverse structural modeling techniques can be employed as well: migration, ray inversion, and map migration.

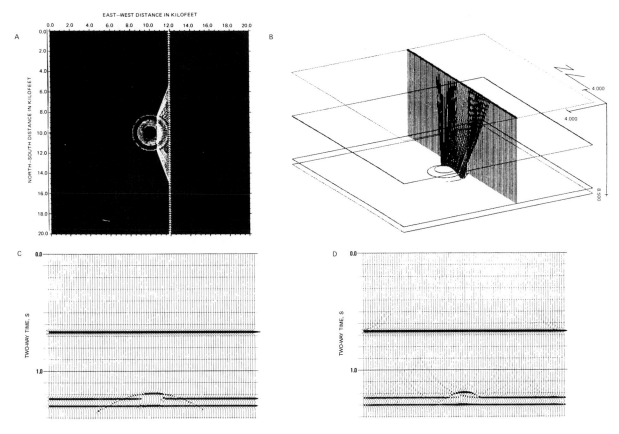

Fig. 11. Contour map with ray paths superimposed (A), 3-D perspective plot (B), unmigrated synthetic seismic section (C), and migrated synthetic seismic section (D) for line shot over the buildup toe-of-slope. Normal incidence rays are captured from high on the flank of the buildup indicating 350 ft (107 m) of buildup thickness where only 50 ft (15 m) is actually present. The unmigrated and migrated seismic section record a significant thickness of buildup due to sideswipe.

208   **Rudolph and Greenlee**

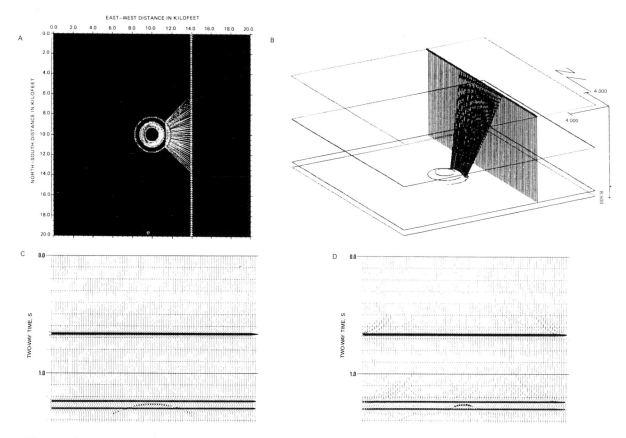

Fig. 12. Contour map with ray paths superimposed (A), 3-D perspective plot (B), unmigrated synthetic seismic section (C), and migrated synthetic seismic section (D) for line shot 2000 ft off the buildup toe-of-slope. Due to sideswipe, an anomaly is present in both unmigrated and migrated seismic sections.

ACKNOWLEDGMENTS

The writers thank Exxon Production Research Company for permission to publish this paper. S. W. Fagin and M. H. Feeley reviewed the manuscript and made many helpful suggestions.

REFERENCES

Caughey, C. A., 1988, Seismic delineation of algal mound reservoirs, Humble City South Field, Lea County, New Mexico, abs., Am. Assn. Petr. Geolog. Bull. v. **72**, p. 170.

Grayston, L. D., Sherium D. F., and Allan J. F., 1964, Middle Devonian in R. G. McCrossan, R. G., and Glaister, R. P., Eds., Geological history of Western Canada: Alberta Soc. Petr. Geol., 49–59.

Kendall, A. C., 1975, The Ashern, Winnepegosis, and lower Prairie Evaporite Formations of the commercial potash areas: *in* Christopher, J. E., and MacDonald, R., Eds., Summary of investigations 1975 by the Saskatchewan Geological Survey: Sask. Dept. Min. Res. Misc. Rept., 61–65.

Neidell, N. S., and Poggiagliolmi, E., 1977, Stratigraphic modeling and interpretation—geophysical principles and techniques, *in* Payton, C. E., Ed., Seismic stratigraphy—Applications to hydrocarbon exploration: Am. Assn. Petro. Geologists Memoir **26**, 389–416.

Perrin, N. A., 1982, Environments of deposition and diagenesis of the Winnepegosis Formation (Middle Devonian), Williston Basin, North Dakota: Proceedings, 4th Internat. Williston Basin Symp., Regina, Sask., 51–66.

CASE HISTORY 9

Defining a Salt Sill Using Three-Dimensional Ray-Trace Modeling and Inversion

*Stuart W. Fagin**

ABSTRACT

Accurate definition of the depth structure of a salt sill, with two-dimensional-seismic data, is made difficult because of sideswipe, raypath bending, and velocity anomalies. A procedure is described, using three-dimensional forward modeling and map migration, which leads the interpreter to a geophysically plausible model.

INTRODUCTION

Complex structures pose difficult imaging problems for several well-known reasons: three-dimensional (3-D) effects, velocity effects, and raypath bending effects. Our imaging procedures often fail to overcome these effects because they require either prior knowledge of structure or 3-D data sets. And yet each two-dimensional (2-D) line represents a sampling, or illumination, of some portion of subsurface structure. Each reflection holds the potential for constraining some aspect of a structural interpretation. The problem, for each reflection, is to determine the geologic surface which has given rise to it, and where on the surface the reflection has come from.

If the reflection points associated with a particular reflection lie within the plane of section, data migration will usually give a reasonable approximation of the position of the surface. If reflection points are located out of the plane of section, i.e., there is sideswipe, data migration will give an erroneous positioning of the reflector. Only the intensive spatial sampling of a 3-D data set will allow for the correct positioning of the data by data migration. However, in the absence of a 3-D data set there is another alternative.

Because an interpreter is able to associate reflections from one line to the next over an unmigrated 2-D grid he can define the 3-D shape of a reflection with less data than that required by a 3-D wave theoretical migration procedure. This definition of 3-D reflection structure when combined with an estimate of subsurface velocities can yield a definition of reflector structure through map migration.

Map migration is often regarded simply as an adjustment made to an existing depth map to account for some geophysical effect, particularly when image-ray

*Exxon Co. U.S.A., Houston, TX.

map migration is employed. Often the adjustment is so slight the effort doesn't seem to be justified. However, this view of map migration is too limited.

When employed on unmigrated data map migration is a powerful interpretational tool. Map migration gives insight into the manner in which reflection events, in all their manifestations, (reflections, sideswipe, and diffractions) are organized in three dimensions, and in so doing allows the interpreter to more confidently account for events on the seismic section. Ultimately, the process will allow the interpreter to derive a 3-D depth solution that is geophysically plausible; a solution capable of recreating observed reflection events on every line in the seismic grid, while properly taking into account the raypath bending, sideswipe, and velocity anomaly problems which hinder complex structure interpretation.

These points are illustrated with the example presented. The technique is practical in that it involves a level of effort suitable for prospect definition and well planning in an exploration setting. The work discussed here required five weeks in which the seismic data were interpreted (both unmigrated and time-migrated lines), the map migration was performed, and each line in the grid was forward modeled. Indeed, the final week involved a second iteration through the entire process where the forward modeling results were used to guide an expanded interpretation of the unmigrated data. However, for the analysis to be swift, the data must be interpreted on an interactive system that allows easy data entry to the map migration and modeling programs.

DESCRIPTION OF THE PROBLEM

The data grid studied consists of eleven lines shown in Figure 1. For each line unmigrated and time-migrated versions were interpreted. Figure 2 shows interpreted and uninterpreted unmigrated sections for the eleven lines. In the following discussion, the salt structure is described from these lines ignoring, for the moment, the consequences of the lack of migration.

The salt structure is located in deep water (2.1–2.8 s) at the base of the present-day continental slope. Local bathymetry reflects recent movements of the underlying salt structure which created gentle to moderate bathymetric gradients.

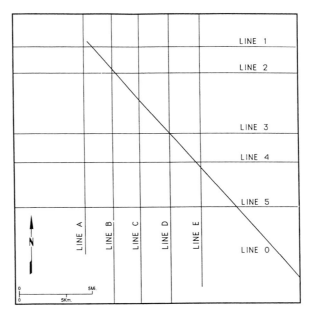

Fig. 1. Base map showing grid of lines used in this study.

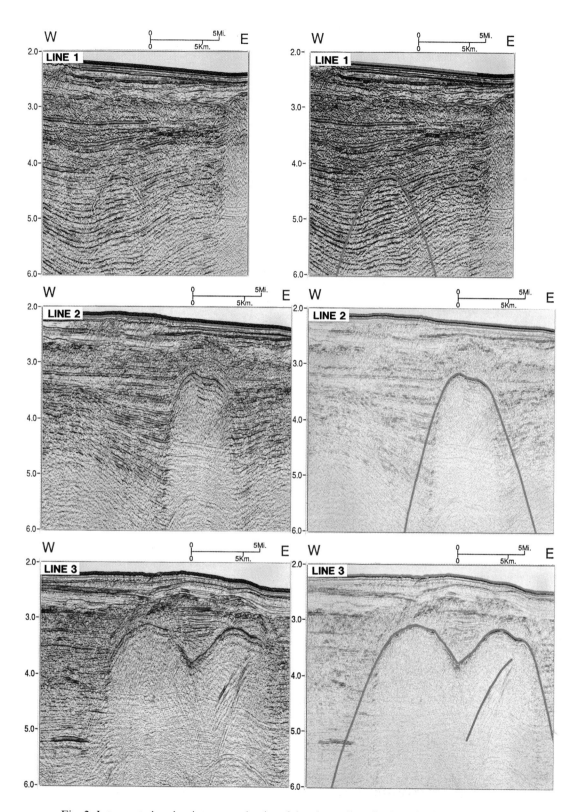

Fig. 2. Interpreted and uninterpreted pairs of the eleven lines in the seismic grid. Top of salt is green, and base of salt red.

Onlap surfaces in the section above salt reveal the timing and depth of emplacement of the salt mass.

Line D, which transects the eastern portion of the structure, shows two distinct
lobes of salt, and an intervening low. The southern lobe is silled and can be seen
to have flowed in a basinward (southern) direction. The base of salt event
underlying this lobe is concave downward and dips steeply to the north, where it
appears to form the flank of a feeder system located beneath the intervening low.
The base-of-salt is highly discordant to the south-dipping subsalt section which
thins northward toward the structure. The northern lobe has no visible base of salt
event and may either be a salt stock or sill. To the west, on Line A, the intervening
low disappears and a single lobe is present.

Because the salt sill is a prominent sealing unit, defining the base-of-salt
structure is an important exploration objective. However, mapping this surface
from the migrated sections is made difficult for several reasons. First, the
base-of-salt event rarely ties from one migrated line to the next. Figure 3 shows an
example of the mis-tie at the intersection of Line 4 and Line D. Because of 2-D
migration effects mis-ties are common when mapping structures with steep dip.
Second, the strong velocity contrast with the enveloping sedimentary section
(6000–8000 ft/s versus 14 700 ft/s) gives rise to velocity pullups and sags. Third,

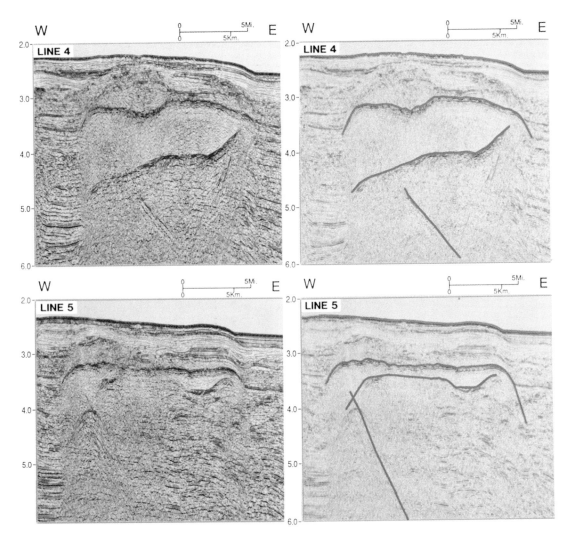

Fig. 2. Cont.

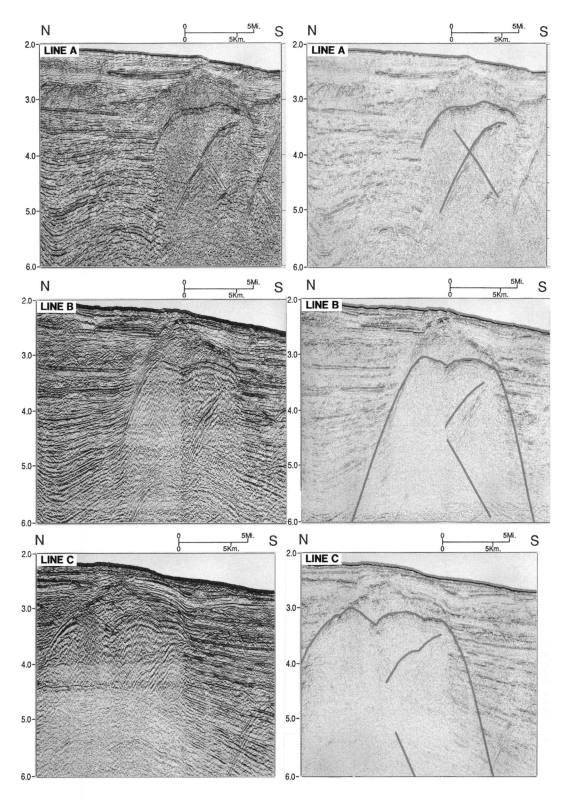

Fig. 2. Cont.

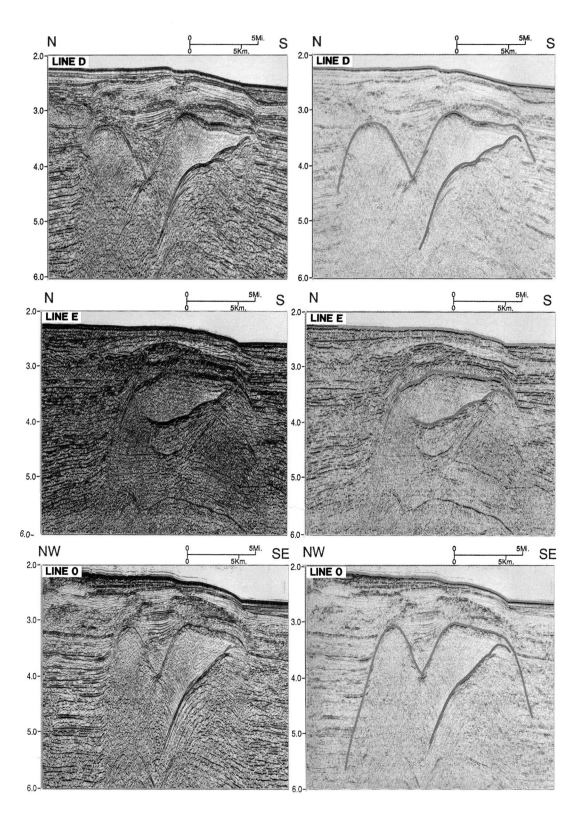

Fig. 2. Cont.

sideswipe can be anticipated on most of the lines given the variable strike and dip of the structure. 2-D migration cannot properly reposition these events. Finally, raypath bending at the top of salt, caused by the strong velocity contrast there, is not well accounted for in time-migration. Map migration, which does not rely on the performance of 2-D wave equation migration, offers a means of avoiding these problems.

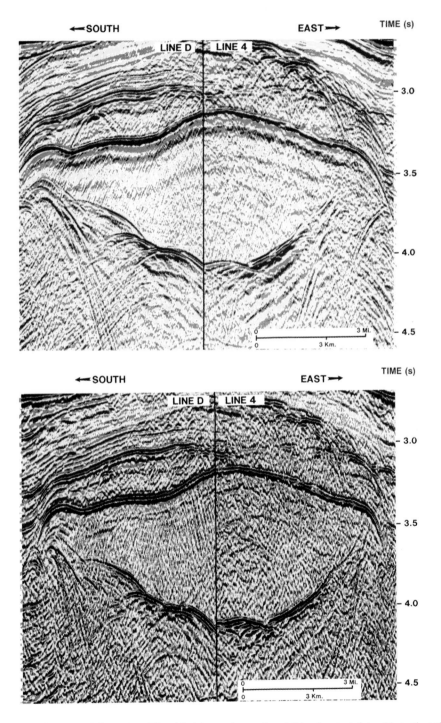

Fig. 3. Tie between Lines 4 and D with (a) unmigrated and (b) migrated data. Note that the base-of-salt reflection ties in the unmigrated grid but not the migrated grid.

CREATING THE INPUT TO MAP MIGRATION: REFLECTION MAPS AND VELOCITIES

Map migration requires two types of inputs: reflection maps of surfaces to migrate and velocities for the intervals between these surfaces. The model developed during migration must include only those surfaces necessary to simulate raypath bending. In the study area only three surfaces were necessary: the water bottom, the top of salt, and the base of salt.

The interval velocities between the surfaces are well constrained. The water layer and salt are 5000 ft/s and 14 700 ft/s, respectively. The section above the salt is only mildly deformed and an interval velocity was easy to define using stacking velocities. The software employed allows for vertical velocity gradients and a reference velocity of 5000 ft/s (at the water bottom) and a vertical gradient of 533 ft/s per 1000 ft were used.

Reflection maps (that is, maps defining unmigrated reflection structure) were made for each of the three surfaces. The construction and migration of the water bottom reflection map involves no interpretational issues and so is not discussed. The top-of-salt and base-of-salt reflection maps are shown in Figures 4 and 5. These maps were made from the interpretation of the unmigrated lines shown in Figure 2. Because the interpretation of these lines requires a different point of view than that taken in constructing depth maps from migrated data, the interpretation is described further.

In constructing depth maps directly from seismic sections, each section is viewed as an image of the earth in cross-section. The interpreter traces reflections with the aim of drawing structures which conform to notions of structural style and patterns for the region. In contrast, in interpreting unmigrated lines, with the aim of constructing reflection maps the seismic section is viewed as a record of recorded reflections according to the zero-offset model. The objective is not to draw a familiar geologic structure but rather to trace out and correlate seismic events.

The difference in the resulting interpretations is significant. For instance, in reflection interpretation, reflections from different surfaces commonly cross (see Line 1). The crossing simply means that some receiver has recorded reflections

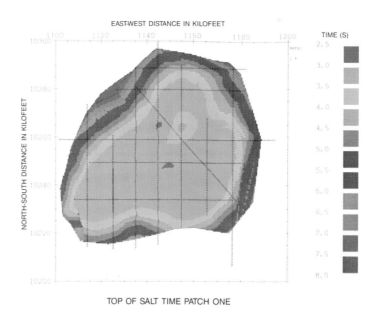

Fig. 4. Map of top-of-salt reflection made from an interpretation of unmigrated data.

from two different surfaces at the same time. In contrast it is unlikely for two geologic surfaces to cross one another. Similarly, in reflection interpretation several reflections may appear from the same surface, for example the familiar bow tie from a syncline. In contrast, in interpreting seismic data as a cross section, surfaces would only repeat if a terrain has been tectonically compressed. This is not to say that the two types of interpretations are totally divorced. The interpreter must have some geologic model in mind, perhaps broadly compatible with the migrated sections, to guide the interpretation of the unmigrated reflections.

Bearing these differences in mind, we can now examine the interpretation of the unmigrated data and the resulting reflection maps. Both the top of salt and base of salt have multiple reflections. The multiple reflections from the top of salt arise from the bow tie associated with median low between the two lobes (see Lines D

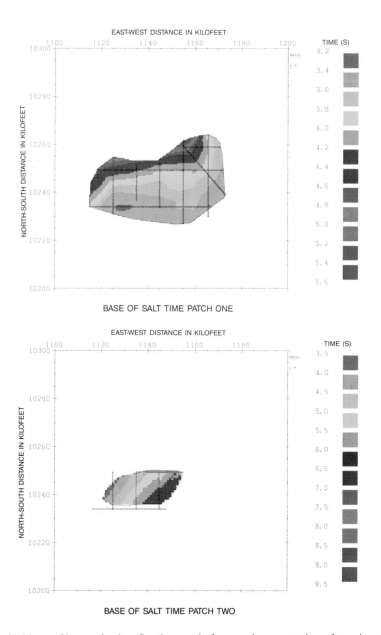

Fig. 5. (a-b) Maps of base-of-salt reflection made from an interpretation of unmigrated data. Two distinct reflections were defined on the unmigrated sections.

and O). The bow tie was handled by creating two separate maps; one for the main reflection and one very limited map for the small basal portion of the bow tie. The purpose of the second map was to maintain some depth control for the bottom of the syncline.

The multiple reflections from the base of salt also arise from a structural low. However, the manifestation of these two separate events is much more subtle. Line 4 shows one very distinct base-of-salt event dipping moderately to the west. A second event, which was ultimately shown by modeling to also be a base-of-salt event, can be seen dipping steeply to the east. As with the top of salt two maps were constructed to represent each base-of-salt reflection. These maps are shown in Figures 5a and 5b. Figure 5c shows the spatial overlap between these two maps. When reflection maps overlap, as in Figure 5c, the interpreter anticipates that upon migration each would be associated with reflection areas that are spatially distinct. If this overlap does not occur then either the velocities used were inappropriate or the association of the two reflections to the same surface is improper.

In further examining the top-of-salt map, an apron of steep dip can be seen to ring the main structure. In comparing this map to the sections, the steep dip is seen to represent the diffraction extension from the top-of-salt reflection. The interpreter would anticipate that upon migration this apron of dip would collapse to a diffraction line source which would define the edge at which the top of salt wraps around and joins with the base of salt. If this does not occur, one would presume some error in the velocities used.

Indeed, because this diffraction collapses to a line, one would not expect to obtain much additional information about subsurface structure by tracing the diffraction to great depths. However, it is useful to trace the diffraction far enough to account for events on other lines. For example, Line 1 contains a distinct event peaking at 4.2 s. However, this event overlaps throughgoing reflections dipping gently to the west, which must represent in place sedimentary section. The nature of this deep event is revealed if the tie between Line 1 and Line D (displayed in Figure 6) is considered. The diffraction emanating from the rim of the north lobe can be seen to tie with the event on Line 1, conclusive evidence that this event is

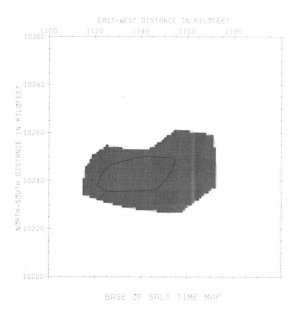

Fig. 5. (c) Map showing superposition of these two reflection segments showing they overlap in unmigrated space.

caused by sideswipe from a structure out of the plane of section. Of course, one might recognize this as sideswipe simply by the appearance of superimposed images. However, sideswipe is not always so blatant and it is important to make these ties and verify these relationships. In addition, sideswipe or not, this event ties with the other events from the top of salt and constitutes valid control for the top-of-salt reflection map.

These reflection maps which represent reflection structure for all the relevant surfaces, and the accompanying velocity information, constitute the sole input to map migration.

MAP MIGRATION: DERIVING THE DEPTH MODEL

Map migration works in a successive manner from the top down. As each surface is defined from its migrated reflection it bends the raypaths used to migrate deeper surfaces.

The migrated water bottom is the first surface in the model. Using the layer velocity information for the above-salt section this model is used to migrate the top of salt. Migration raypaths (connecting recording points along the ground surface with reflection points) to the top of salt are traced. Figure 7 shows the depth points defined by the ends of these rays. As would be expected the vectors associated with the diffraction apron have collapsed to define the outer rim of the salt structure. Grid points which represent evenly spaced sampling of the reflection surface are collapsed to a dense line which tightly defines the edge of salt. Each of these grid points contains depth information which acts as control for surface definition. In practice, the interpreter needs to edit these points to remove spurious results.

The top-of-salt surface, when defined, is combined with the water bottom surface to form a model through which the base of salt reflection maps may be migrated. Figure 8 shows the migrated points associated with the base of salt. Where the reflection maps spatially overlapped, the control points can now be

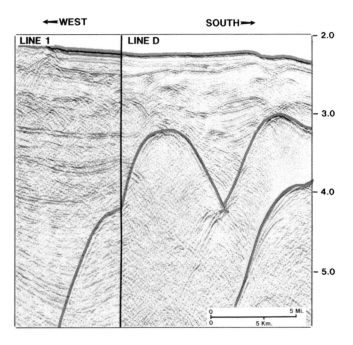

Fig. 6. A tie between unmigrated Lines 1 and D. Note that the top-of-salt sideswipe event from Line 1 ties to the diffraction on Line D.

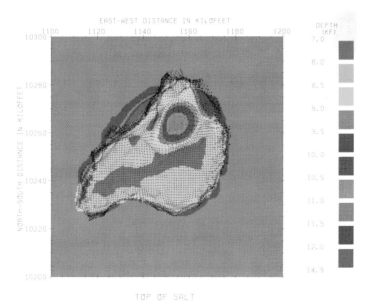

Fig. 7. The depth points associated with the top-of-salt migration vectors and the derived top-of-salt structure map. This grid (which is uniformly spaced over the unmigrated reflection map) has collapsed to define the diffraction's source at the rim of the salt sill. The structure contour map upon which these points are superimposed was constructed from the depth information associated with these points and from user-added points where control was lacking.

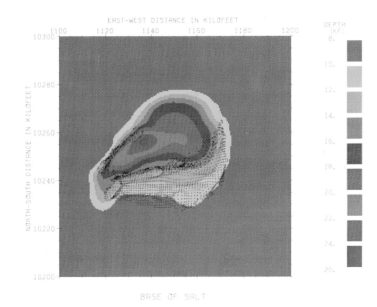

Fig. 8. The depth points associated with the base-of-salt migration vectors and the derived base-of-salt structure map. Note that the two reflection maps are associated with spatially distinct reflection areas on the base-of-salt surface. The larger map (Figure 5a) is a reflection from the southern flank of the base-of-salt and the smaller map (Figure 5b) from the northwestern flank.

seen to define separate portions of the base-of-salt structure. The base of salt is shown to be a triangular shaped structure with an elongate northeast-trending feeder. One set of control points defines the northwest flank, one set the southern flank, and the northeastern flank remains undefined. Before final gridding a series of control points is added to the northeast flank to complete the structure in a geologically reasonable way.

FORWARD MODELING: RECREATING THE UNMIGRATED SECTIONS

The base-of-salt surface, when completed, is combined with the top of salt and water bottom to form the final model. At this point the analysis can be considered completed and these depth maps regarded as the final product. Alternatively, one can use the depth model for 3-D forward ray-tracing to see if the forward modeling recreates the unmigrated sections.

At first glance this would not seem to be a useful exercise as one would be tracing the same rays and, therefore, obtaining the same results without gaining additional insight into the structure. In fact, there are several good reasons for performing the forward modeling. First, contained in the map migration step is the need to delete and add points to the migrated control, and perhaps smooth the resulting grids. In performing these steps error can be introduced in a way that can only be revealed by forward modeling.

Second, as already alluded to, the results of forward modeling act as a very effective guide to interpreting the unmigrated sections. Events whose significance or nature were not previously recognized may be better understood with the benefit of modeling results. After modeling, the unmigrated interpretation could be expanded to include more events and presumably more depth control after migration. The steep east-dipping event on Line 4 was not initially interpreted as a base-of-salt event. Only after modeling was it realized that a base-of-salt event of similar dip and in a similar position as this event should be expected.

Third, forward modeling allows the interpreter to gain a better understanding of the control on depth structure. Forward modeling allows the interpreter to understand where the model could be altered, pursuant perhaps to geologic constraints, while still maintaining compatibility with geophysical observations.

With these motivations in mind the depth model was forward ray-traced along each of the lines in the grid to give the results shown in Figures 9 through 19. For each seismic line the modeled arrival times and interpreted sections are shown side-by-side. Examination of these comparisons shows that the model, with few exceptions, effectively recreates the unmigrated events and must be regarded as a highly plausible model.

Figures 9 through 19 also show perspective and map views of the ray-tracing to both the top and base of salt for each line. For many of the lines these displays clearly depict sideswipe, particularly with regard to the base of salt. Displays (g) and (h) juxtapose time-migrated images against time-scaled cross sections through the model. In Figure 20, these two sections are superimposed for each line. These illustrations compare the structure defined by ray tracing (time-scaled) and the structure imaged by time-migration.

These displays call attention to several aspects of the seismic response of this structure which would be difficult to recognize without the benefit of modeling. First, many events which would appear to be subsalt events are, in reality, base-of-salt reflections. The steep, east-dipping event on Line 4 was previously mentioned. Another example is the event on Line A at 4.7–4.8 s at the bottom of the interpreted south-dipping base-of-salt event. The synthetic result indicates this event may be an additional part of the base-of-salt reflection. The impact of

LINE 1

W   E

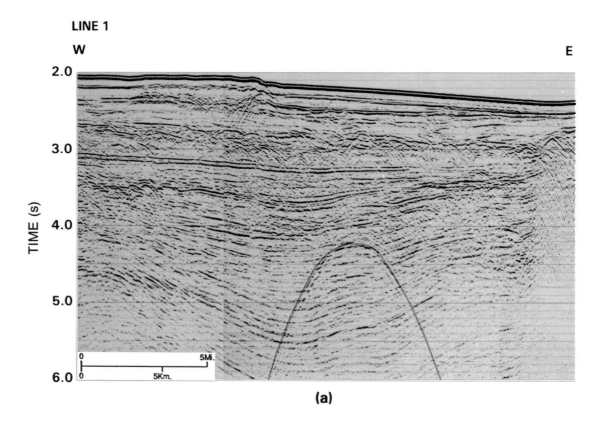

(a)

LINE 1

W   E

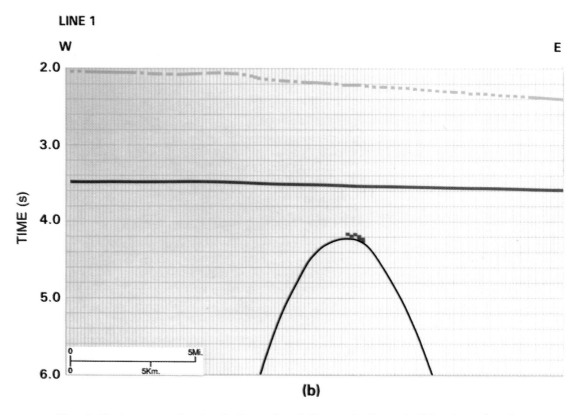

(b)

Figs. 9–19. A montage showing the forward modeling results for each of the eleven lines in the grid. (a) The unmigrated sections with interpretation (top of salt is in green, base of salt is red); (b) synthetic arrival times with the original interpretation shown superimposed in black; (c–d) map view and perspective views of the top-of-salt reflection raypaths; (e–f) map view and perspective view of the base-of-salt raypaths; (g) a time migrated section; (h) a time-scaled cross-section through the model along the plane of the seismic line.

222

these interpretational insights can be great if subsalt exploration is contemplated. In addition, numerous differences can be seen between the structure one would define from the time-migrated grid and that defined by ray tracing. In particular, note the differences in the top-of-salt event in Figure 20 on Lines 3, A, and C and the base-of-salt event on Lines 4 and E.

These figures constitute the most complete definition of the structure that can be achieved from the seismic data. All of the observable reflections from the top and base of salt are well simulated. This sort of match is difficult to achieve in three dimensions by forward modeling alone, because the numerous modifications to structure and modeling iterations in the procedure would be too arduous.

SUMMARY

Many reviews of complex structure seismic interpretation begin with the disclaimer that the analysis is subject to numerous sources of error such as those mentioned in the opening paragraph. The result is often to discredit the prospect

(continued on page 243)

LINE 1

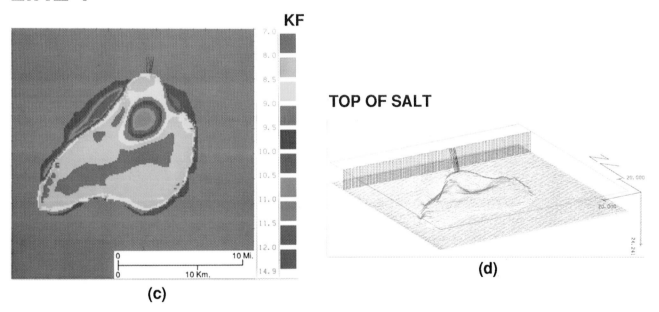

(c)

(d)

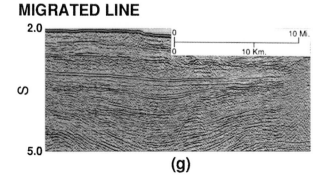

(g)

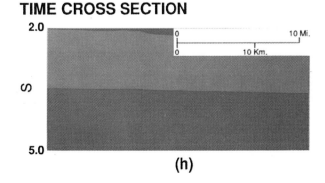

(h)

Fig. 9. Cont.

LINE 2

W

E

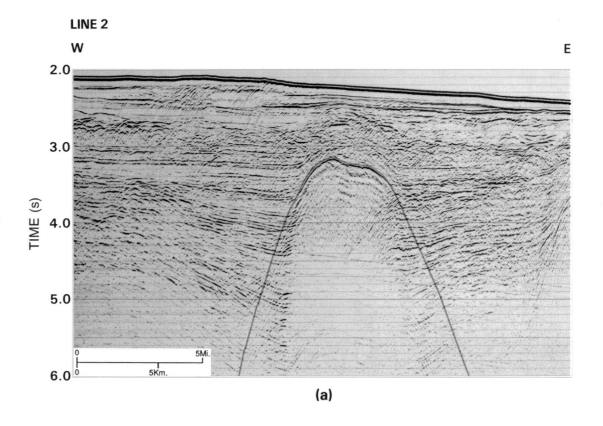

(a)

LINE 2

W

E

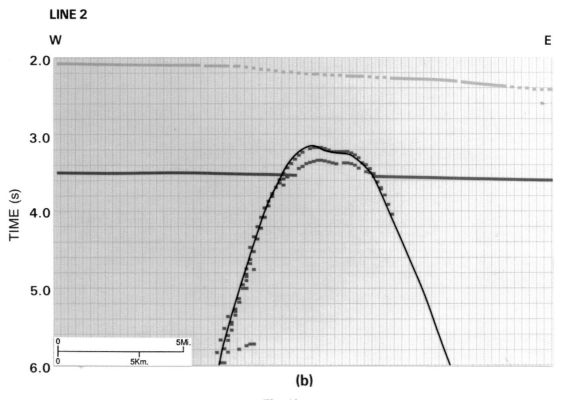

(b)

Fig. 10.

LINE 2

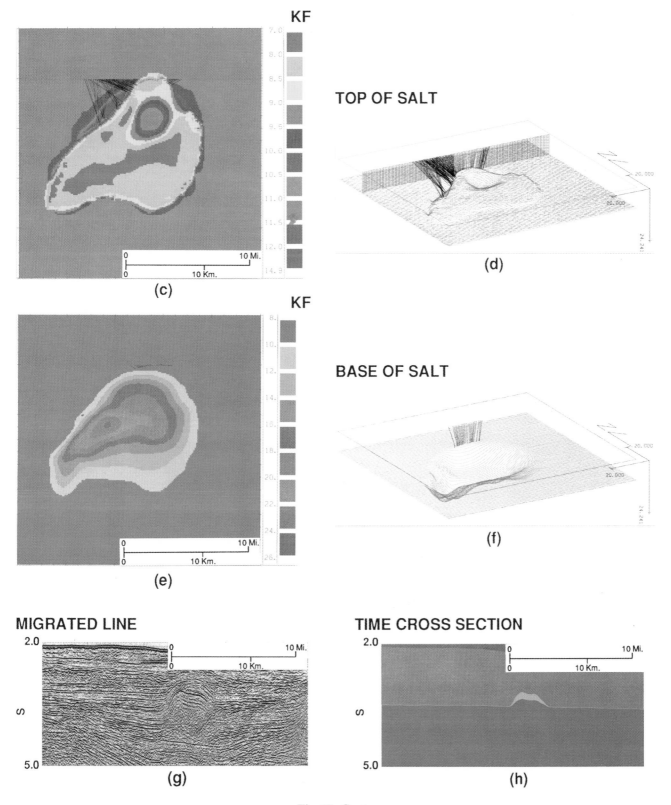

KF

7.0
8.0
8.5
9.0
9.5
10.0
10.5
11.0
11.5
12.0
14.9

0   10 Mi.
0   10 Km.

(c)

TOP OF SALT

(d)

KF

8.
10.
12.
14.
16.
18.
20.
22.
24.
26.

0   10 Mi.
0   10 Km.

(e)

BASE OF SALT

(f)

MIGRATED LINE

2.0   0   10 Mi.
                                 0   10 Km.

S

5.0
(g)

TIME CROSS SECTION

2.0   0   10 Mi.
                                 0   10 Km.

S

5.0
(h)

Fig. 10. Cont.

LINE 3

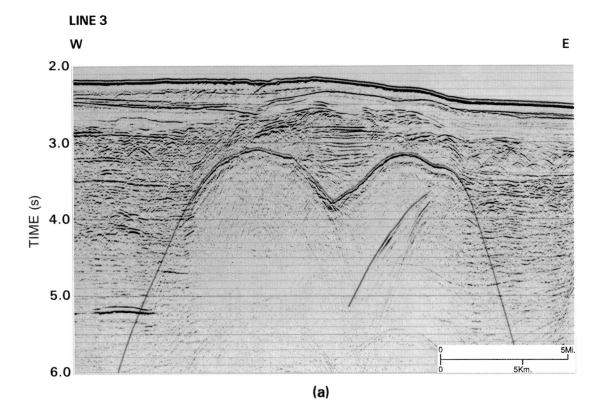

(a)

LINE 3

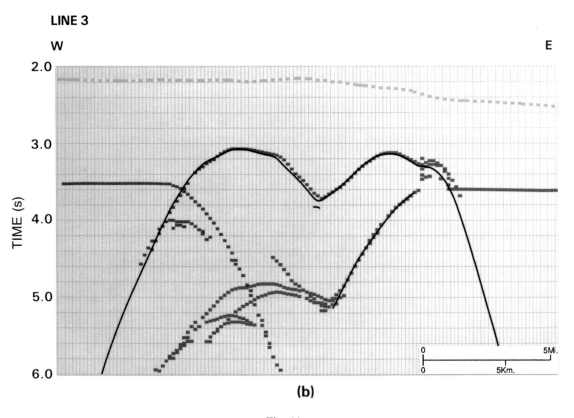

(b)

Fig. 11.

LINE 3

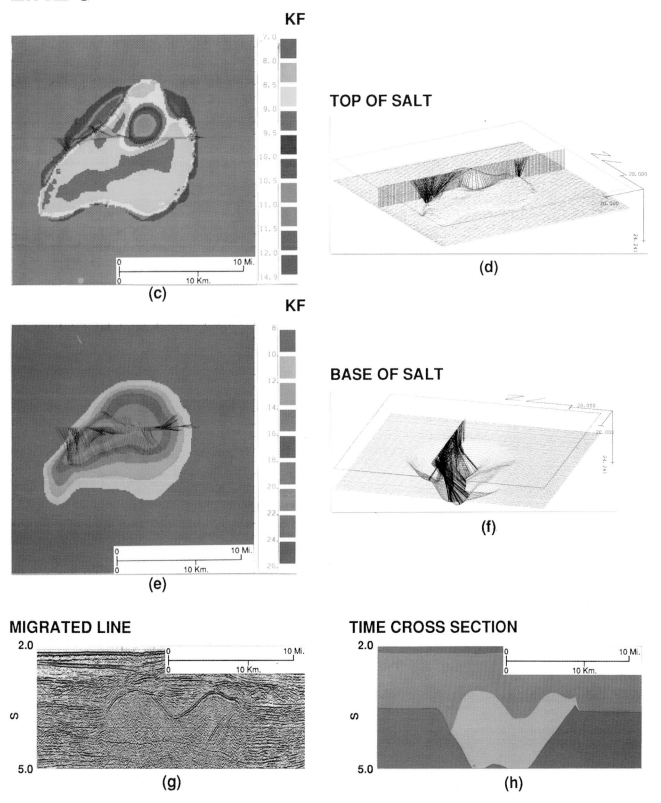

(c)

(d) TOP OF SALT

(e)

(f) BASE OF SALT

MIGRATED LINE (g)

TIME CROSS SECTION (h)

Fig. 11. Cont.

LINE 4

W   E

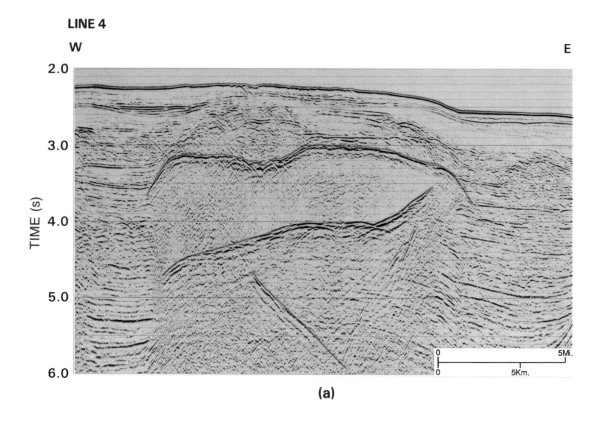

(a)

LINE 4

W   E

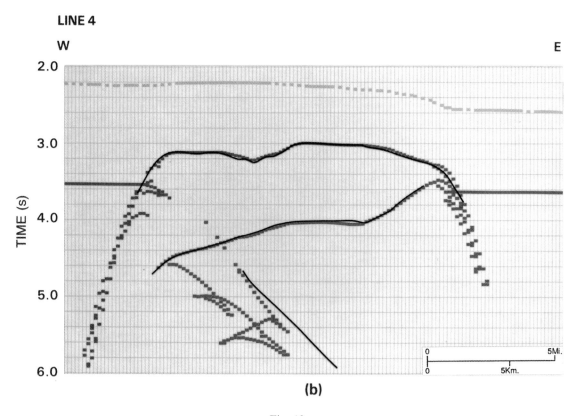

(b)

Fig. 12.

LINE 4

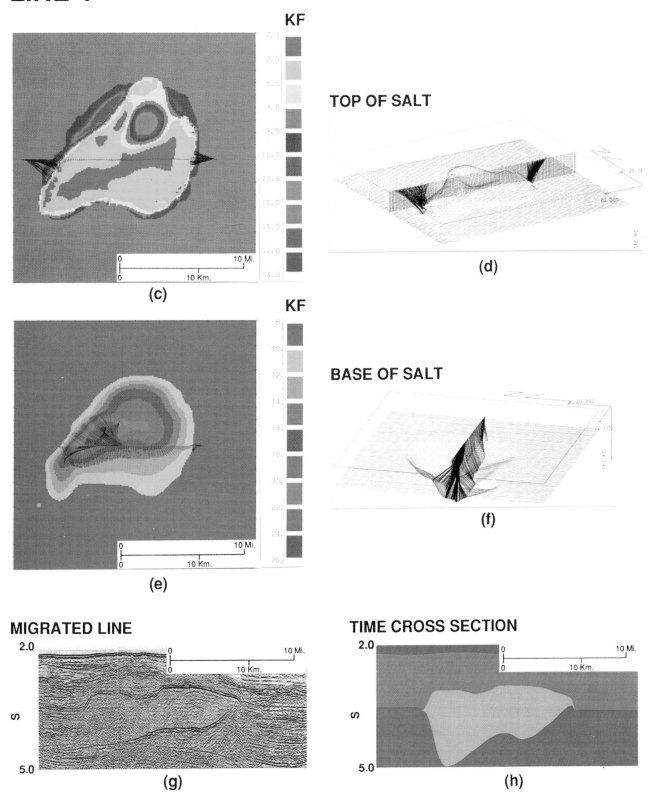

Fig. 12. Cont.

LINE 5

W   E

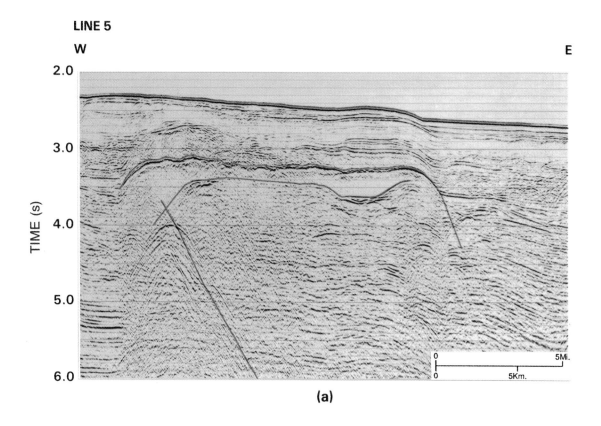

(a)

LINE 5

W   E

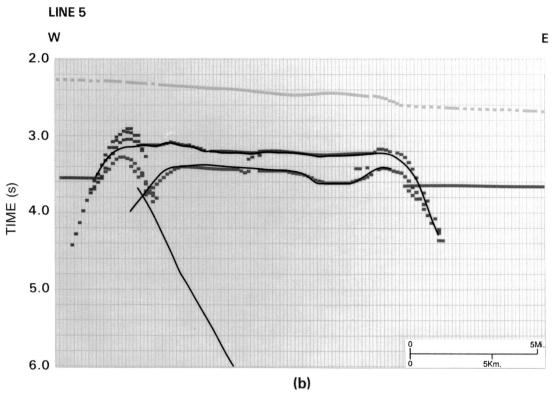

(b)

Fig. 13.

LINE 5

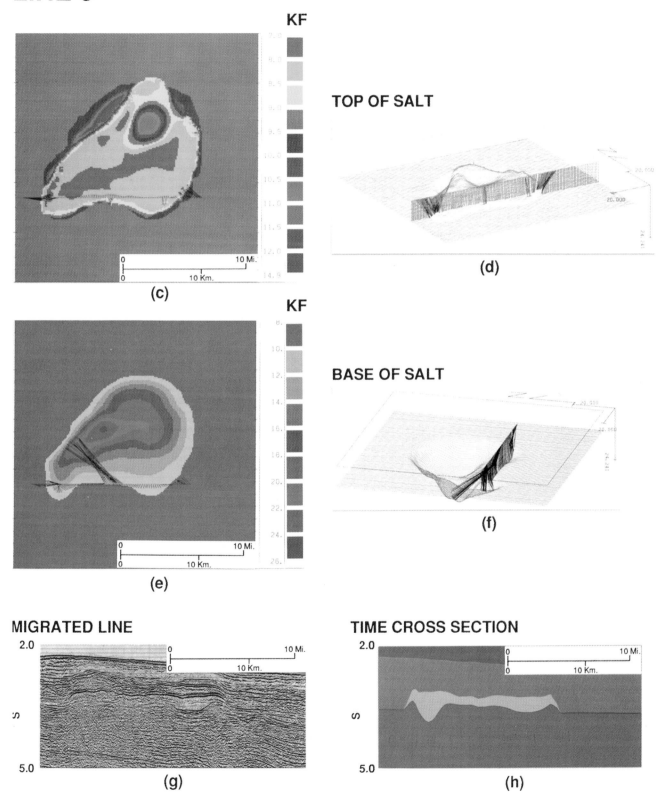

KF

TOP OF SALT

(d)

(c)

KF

BASE OF SALT

(f)

(e)

MIGRATED LINE

(g)

TIME CROSS SECTION

(h)

Fig. 13. Cont.

LINE A

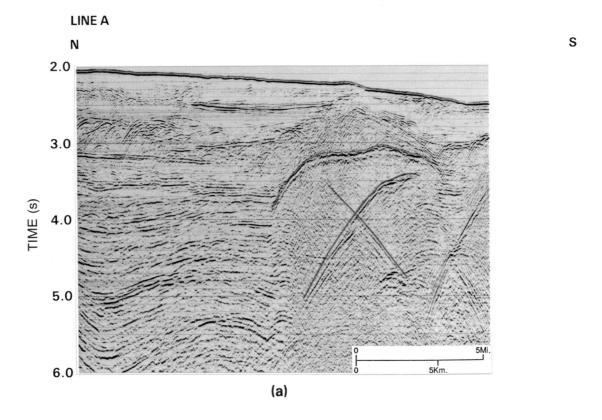

(a)

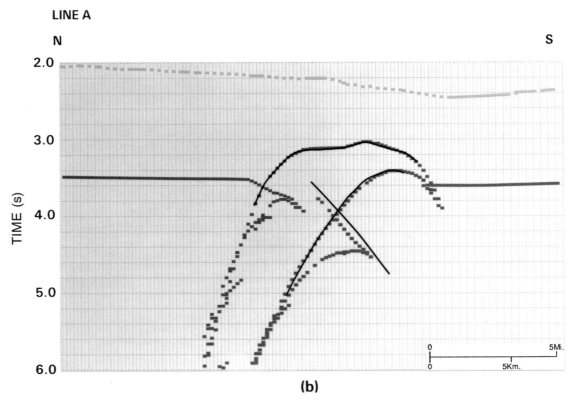

(b)

Fig. 14.

LINE A

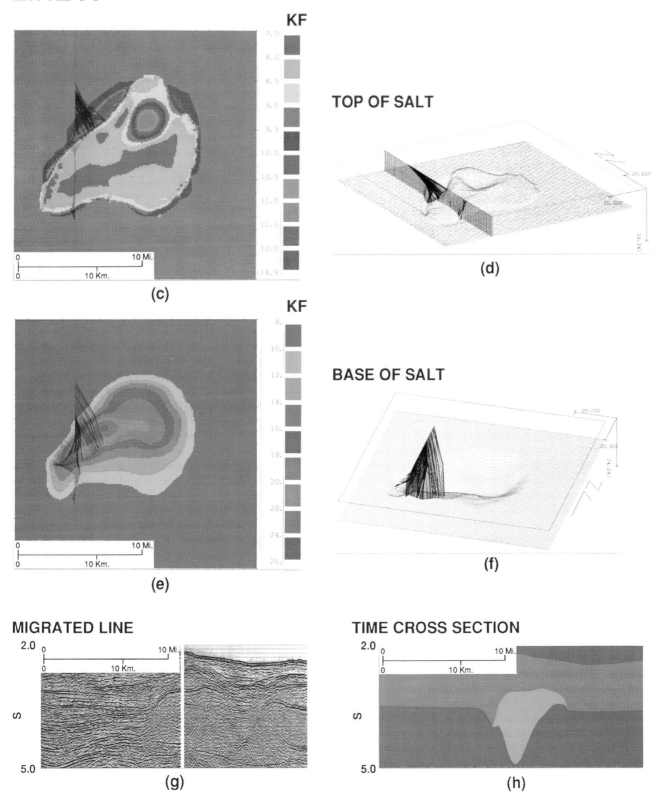

(c)

KF

TOP OF SALT

(d)

KF

BASE OF SALT

(e)

(f)

MIGRATED LINE

(g)

TIME CROSS SECTION

(h)

Fig. 14. Cont.

LINE B

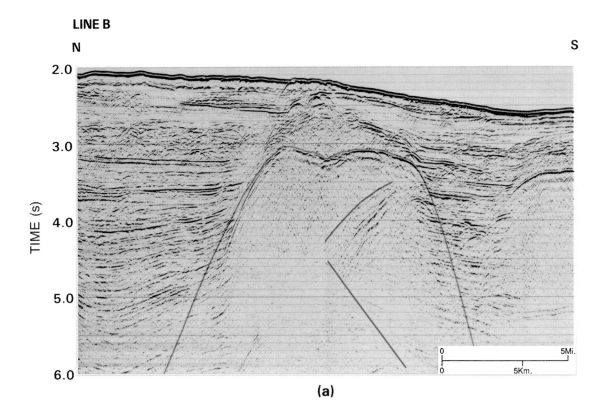

(a)

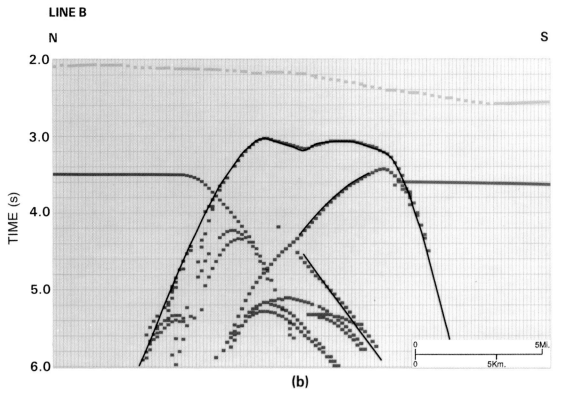

(b)

Fig. 15.

LINE B

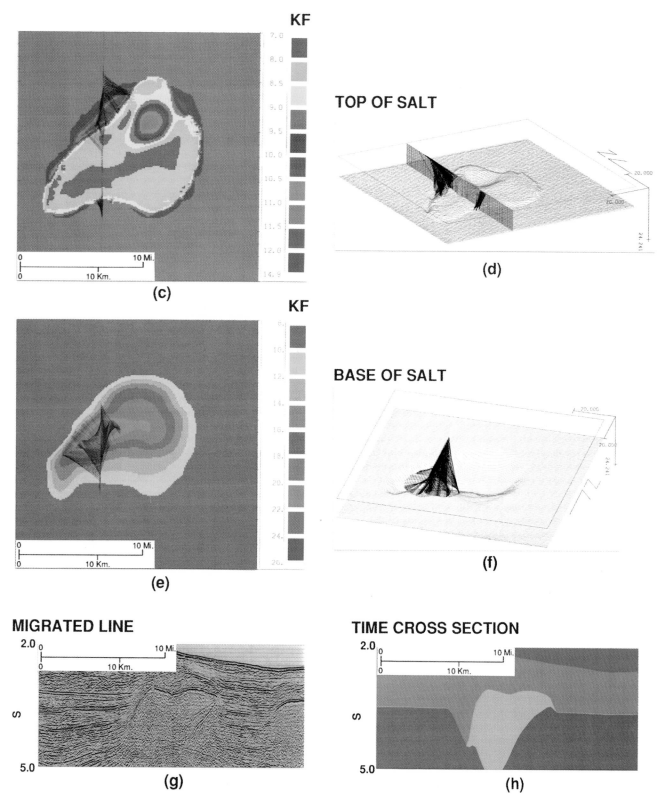

Fig. 15. Cont.

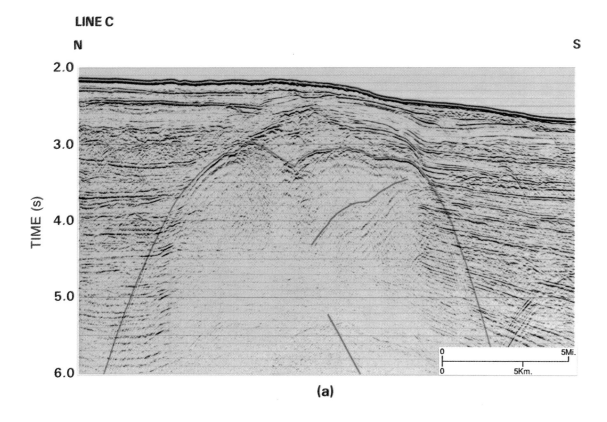

(a)

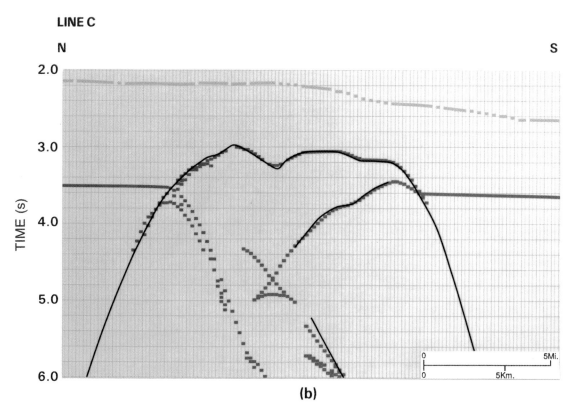

(b)

Fig. 16.

LINE C

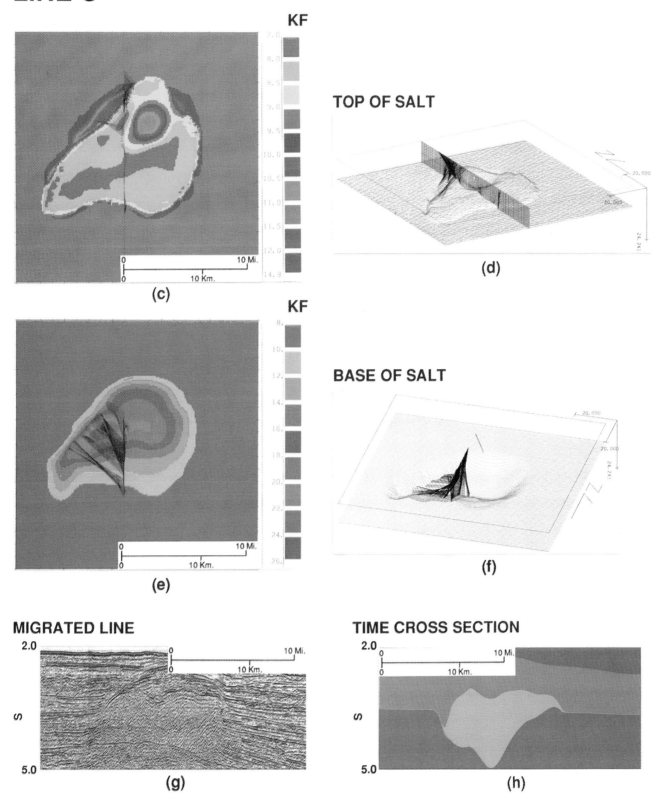

KF

7.0
6.0
8.5
9.0
9.5
10.0
10.5
11.0
11.5
12.0
14.9

0   10 Mi.

0   10 Km.

(c)

TOP OF SALT

(d)

KF

8.
10.
12.
14.
16.
18.
20.
22.
24.
26.

0   10 Mi.

0   10 Km.

(e)

BASE OF SALT

(f)

MIGRATED LINE

2.0

S

5.0

(g)

TIME CROSS SECTION

2.0

S

5.0

(h)

Fig. 16. Cont.

LINE D

N   S

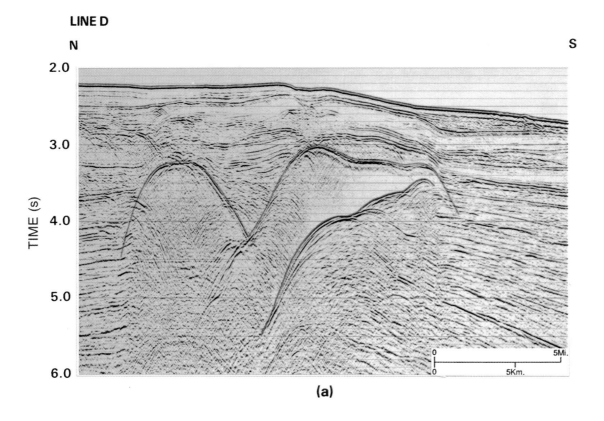

(a)

LINE D

N   S

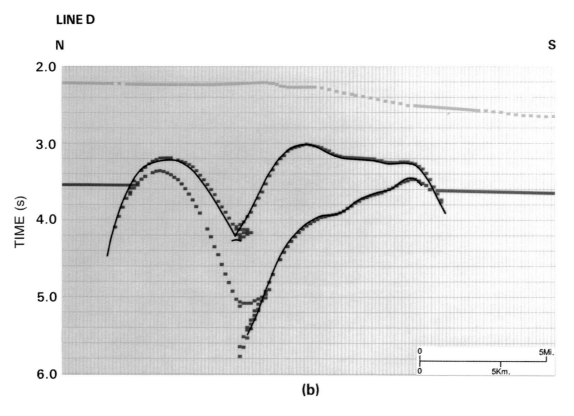

(b)

Fig. 17.

LINE D

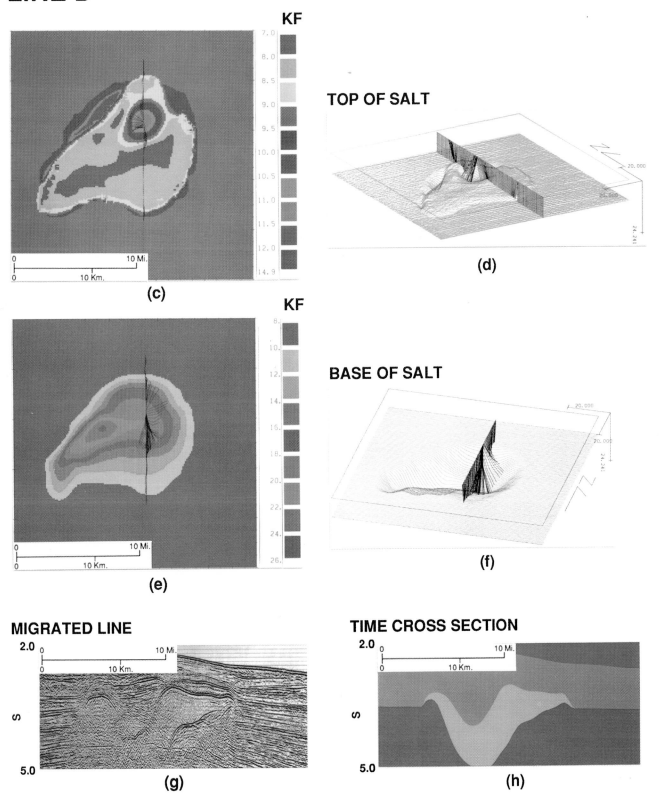

Fig. 17. Cont.

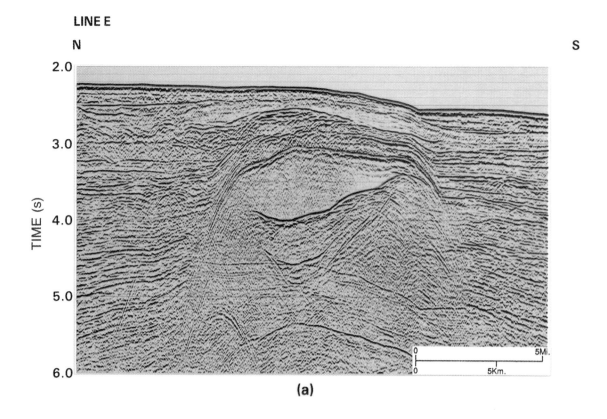

(a)

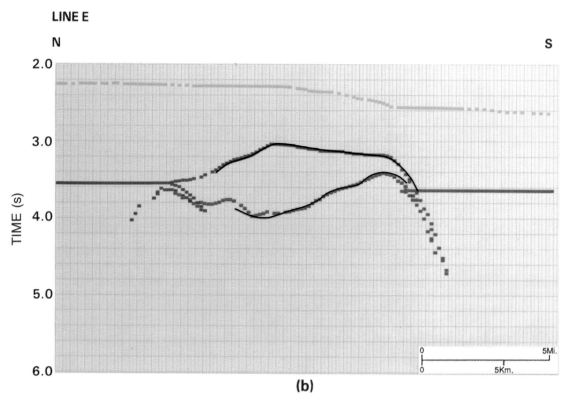

(b)

Fig. 18.

LINE E

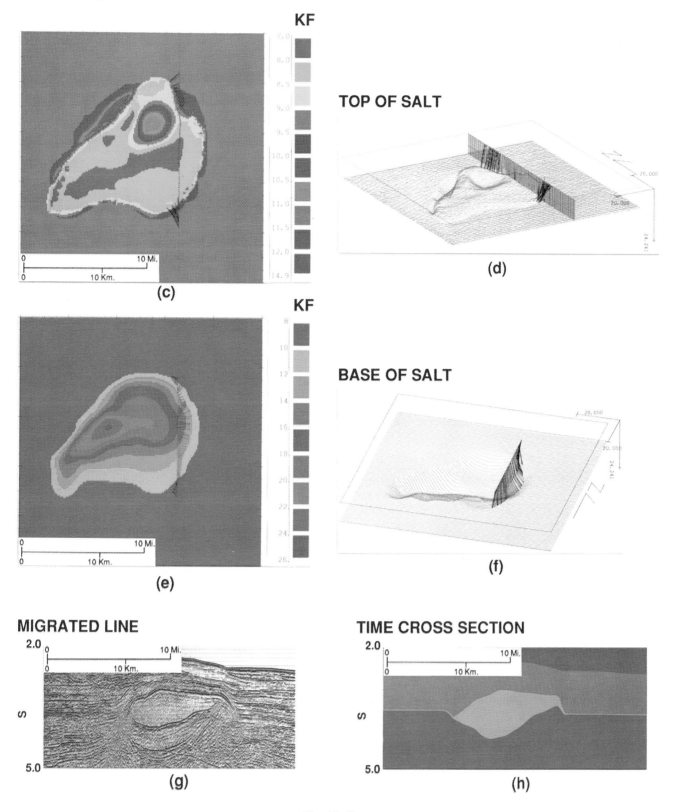

(c)

(d) TOP OF SALT

(e)

(f) BASE OF SALT

MIGRATED LINE (g)

TIME CROSS SECTION (h)

Fig. 18. Cont.

LINE 0

NW

SE

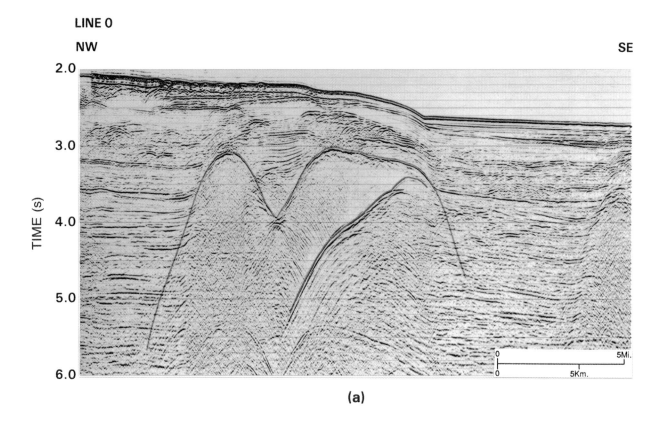

(a)

LINE 0

NW

SE

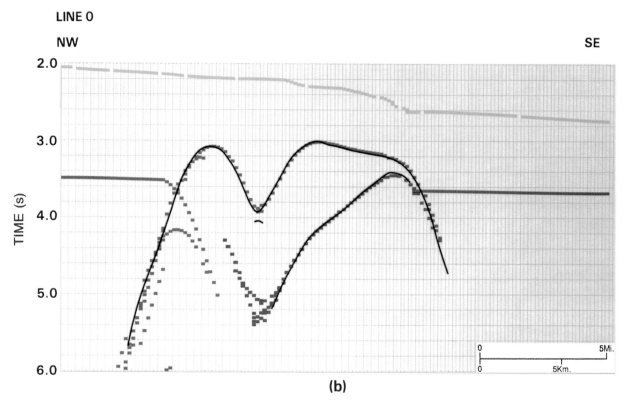

(b)

Fig. 19.

at the outset. Map migration eliminates many of these sources of error and allows for greater confidence in the final interpretation. Only a few sources of error remain.

Velocities are always a source of some error. In this study the velocities were well constrained because of the nature of the units. Most of the length of the raypath to the base of salt traverses the water layer and the salt layer, both of which have extremely well constrained interval velocities (subsequent model-based analysis of moveout patterns from the top and base of salt indicate that salt interval velocities vary no more than a few percent from the value chosen). The section between the top of salt and the water bottom exhibits the most variable interval velocity but analysis of moveout patterns indicate that, to within 5 percent velocities are represented by the vertical gradient that was used. In cases where velocities cannot be so easily constrained, it is best to model end cases within a range of plausible velocities.

(continued on page 248)

LINE 0

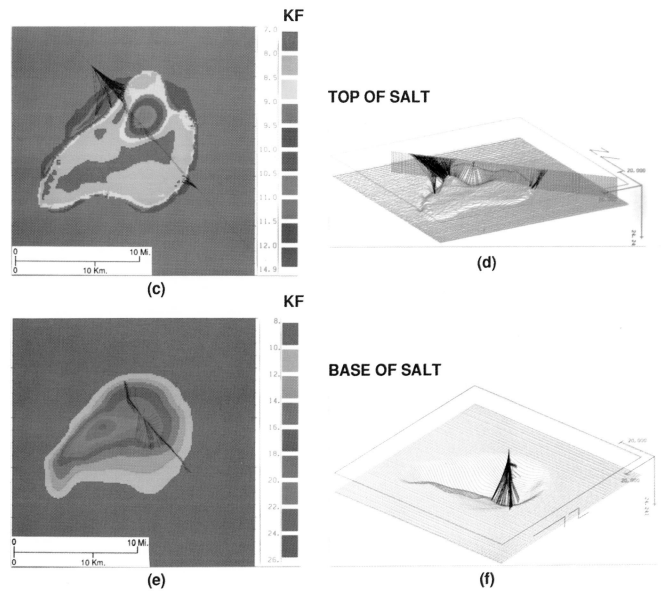

Fig. 19. Cont.

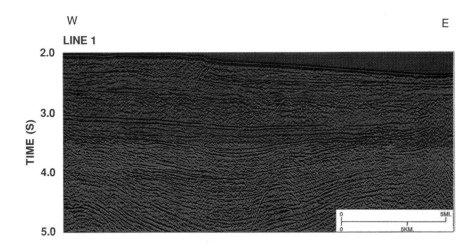

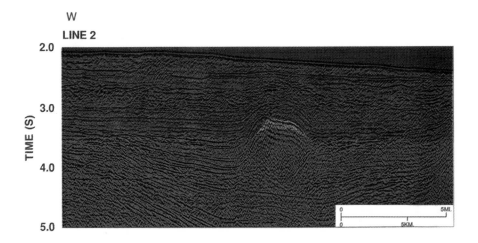

Fig. 20. Superposition of displays g and h from Figures 9–19.

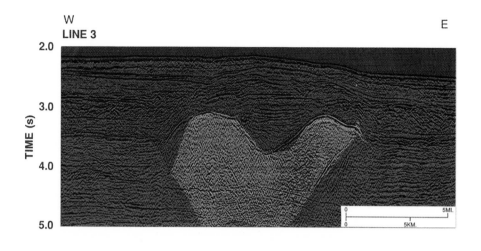

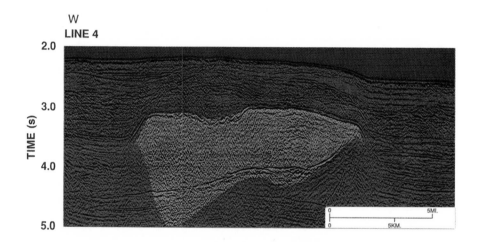

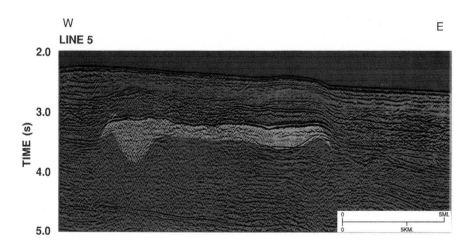

Fig. 20. Cont.

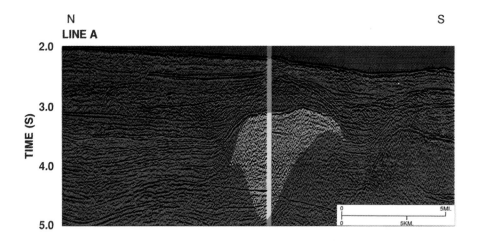

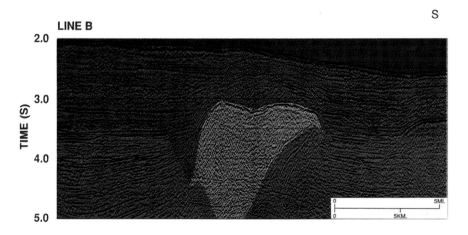

Fig. 20. Cont.

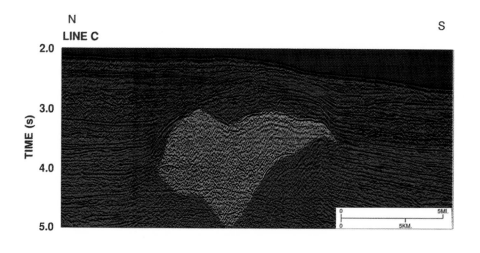

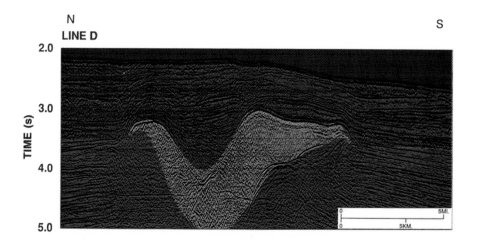

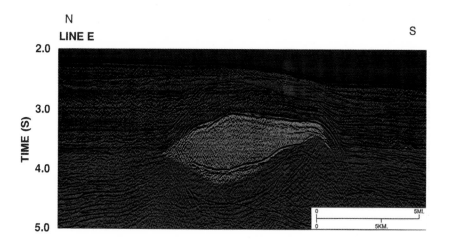

Fig. 20. Cont.

Often the definition of the reflection event itself is the greatest barrier to accurate structural definition. In this study all three surfaces were distinctive and usually there was little doubt as to their nature. Where doubt exists, forward modeling offers the best means of investigating the different possibilities. As was previously described, successive looping between forward and inverse analyses offers the most insightful route to interpreting seismic data properly.

Finally, throughout the analysis the assumption is made that the unmigrated stacked section adheres to the CDP assumption; that stacking the data results in a section that would be obtained by zero offset acquisition. This source of error is always difficult to analyze and, indeed, affects most depth mapping procedures. These errors, arising, for example, from non-hyperbolic moveout, will likely cause loss of signal rather than erroneous zero-offset arrival time values.

CASE HISTORY 10

Integrated Interpretation, 3-D Map Migration and VSP Modeling Project, Northern U.K. Southern Gas Basin*

J. M. Reilly[‡]

ABSTRACT

Depth conversion in the northern portion of the U.K. Southern Gas Basin is complicated by the presence of (Permian) Zechstein salt swells and diapirs. In addition, the post-Zechstein (post-Permian) section displays large lateral velocity variations. The primary agents which control the velocity of this stratigraphic section are (1) depth of burial dependency, (2) lithologic variation within individual formations, and (3) the effects of subsequent tectonic inversion. An integrated approach which combines well velocity, seismic velocity, and seismic interpretation is required for accurate depth estimation.

In 1988 Mobil and partners drilled an exploratory well in the northern portion of the U.K. Southern Gas Basin. This well was located near the crest of a Zechstein salt diapir. Over 2000 m of Zechstein were encountered in the well. The Permian Rotliegendes objective was penetrated at a depth of over 3700 m.

The initial delineation of the objective structure was based on the results of three-dimensional (3-D) map migration of the seismic time interpretation. Spatially variant interval velocity functions were used to depth convert through five of the six mapped horizons. Both well- and model-based seismic interval velocity analysis information were used to construct these functions.

A moving source well seismic survey was conducted. The survey was run in two critical directions. In conjunction with presurvey modeling, it was possible to immediately confirm the structural configuration as mapped out to a distance of seven kilometers from the well. Post-survey 3-D map migration and modeling was employed to further refine the structural interpretation. Although the question of stratigraphic anisotropy was considered in the evaluation of the long offset modeling, no evidence was found in the field data to support a significant effect.

Finally, comparisons were made of: curved ray versus straight ray migration/modeling, midpoint-depth velocity versus (depth dependent) instantaneous velocity functions, and Hubral versus Fermat based map depth migration algorithms. Significant differences in the results were observed for structural dips exceeding 15 degrees and/or offsets exceeding 6 km. Map depth migration algorithms which employed both curved rays and spatially variant instantaneous velocity functions

*Published with the permission of the European Association of Exploration Geophysicists and Geophysical Prospecting.
[‡]Mobil North Sea Limited.

were found to best approximate the "true" geologic velocity field in the study area.

INTRODUCTION

The study area is located in the northern portion of the U.K. Southern Gas Basin (Figure 1). This is an area of active exploration drilling as a result of recent U.K. licensing round awards. Primary targets are (Permian) Rotliegendes and (Carboniferous) Westphalian Sands. A structural cross-section through the primary area of interest which illustrates the basic stratigraphy and structure of the region is shown in Figure 2. The local structure is primarily controlled by Zechstein salt swells and diapirs. These salt features determine both the structure and interval velocity profiles of the overlying post-Permian section. The time structure of the underlying units is dominated by velocity pullup under the diapiric features. The variation in the interval velocities of the major mapping units, as determined from available well and seismic information, is also shown in Figure 2.

The Zechstein interval velocity is dependent on the relative percentages of halite, anhydrite, dolomite, and other minor facies within the section. The lateral variation in the thicknesses of these lithologies within the diapirs and salt swells complicates the interval velocity field.

The Tertiary section displays a velocity variation of over 570 m/s, the Cretaceous Chalk a velocity variation exceeding 1960 m/s, and the Triassic sedimentary section a velocity variation of over 1145 m/s. The velocity within each of these intervals is dominated by the maximum depth of burial (compaction) effect on these sediments (Banik, 1984; Bulat and Stoker, 1987; Carter, 1987). This effect is in turn modified by lateral velocity variations due to stratigraphic factors and post-Cretaceous differential uplift (non-rebound) effects. The average velocity field is further complicated by a velocity inversion, in the synclines, at the base of the Cretaceous Chalk interval.

The migrated depth map at the top of the Zechstein formation is shown in Figure 3. As discussed, the area is dominated by salt swells and diapirs. The total

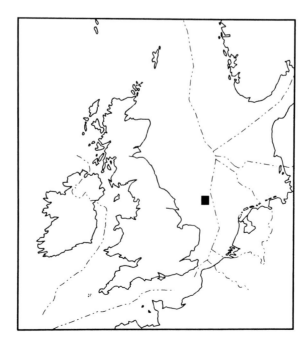

Fig. 1. Location map with the study area shown as a shaded box.

mapping area is over 1600 square kilometers. A stacked (unmigrated) time seismic section is shown in Figure 4. When dip of the section does not exceed 25 degrees, the seismic data quality of all the major mapping horizons (and velocity boundaries) is generally good to excellent.

The stacked (unmigrated) time map at the Base Zechstein/Top Rotliegendes level is dominated by pullup due to the overlying salt diapirs (Figure 5). This results in over 400 ms of time structure underneath the salt swell of interest. Over 900 ms of time pullup is present in other areas of the map. The objective, of course, is to transform the interpreted time surfaces to as accurate as possible, geologically reasonable, structural depth surfaces.

The migrated depth map at Base Zechstein/Top Rotliegendes level is shown in Figure 6. After depth conversion, the structural trend is seen to be predominantly northwest trending, consistent with the interpreted Rotliegendes fault pattern. This trend is in contrast to the northeast trending structures seen on the (unmigrated) time map (Figure 5).

VELOCITY ANALYSIS, VELOCITY FUNCTION DERIVATION, AND MAP MIGRATION

During the initial phase of this project severe problems were encountered in the interpretation of the available migrated time seismic sections. The primary sources of error were the presence of significant off-line dip and event mispositioning beneath the salt diapirs associated with time-domain migration algorithm employed in the seismic processing. As a result, we decided to interpret the stacked (unmigrated) sections and perform 3-D map migration of the time surfaces.

For each of the mapped horizons well and seismic interval velocity information, both from within the area and regional studies, were integrated to produce spatially variant interval velocity fields. These were then used in the "layer cake" vertical depth conversion and 3-D map migration of the area. The primary

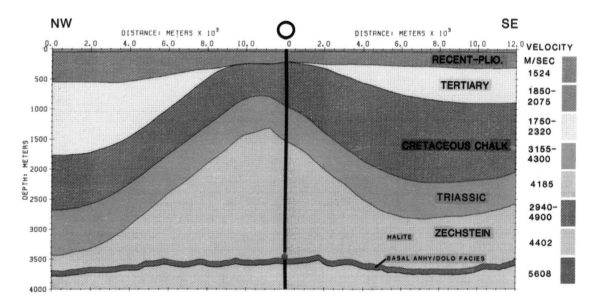

Fig. 2. Structural section through the primary area of interest. Variation in the interval velocities of the major mapping units is shown on the right. Location of the cross section is shown as a solid red line on Figures 3, 5, and 6. The 49/2-3 well location is shown as a black circle. The seismic survey geophone location is shown as a red circle in the well bore at the top of the basal anhydrite/dolomite facies. Vertical exaggeration is three times.

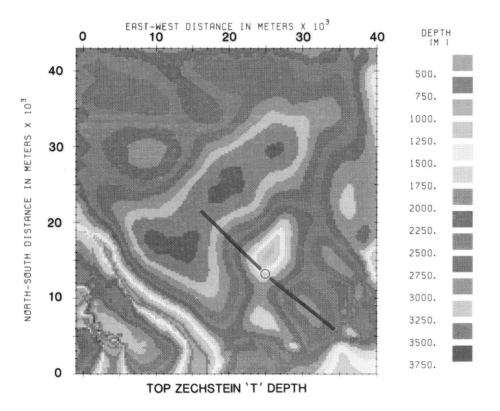

Fig. 3. Migrated 3-D depth map, Top Zechstein, of the interpreted seismic data shows the major salt features within the study area. The Base Zechstein/Top Rotliegendes structure of interest underlies the semicircular salt swell in the right-center of the map. Note the major bounding synclines and the northwest trending salt walls in the southwestern portion of the mapping area. The solid red line is the location of the structural cross section shown in Figure 2. The red circle is the location of the 49/2-3 well. Nine other wells are located within the study area (not shown).

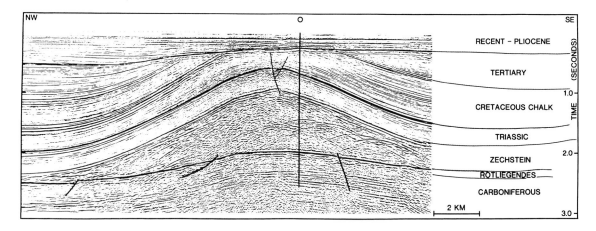

Fig. 4. Stacked (unmigrated) seismic section as an example of seismic data quality in the area. Note the erosional truncation of the Cretaceous chalk in the vicinity of the salt swell. Location of section is coincident with the structural cross section shown in Figure 2. Location of the 49/2-3 well is shown as a black circle.

advantage of well interval velocity information is that, in the absence of dip, the calculated interval velocities can (hopefully) be considered to be "exact". Therefore, the interpreted seismic time horizons can be made to tie exactly in depth to the wells on an interval-by-interval basis. The primary weakness in utilization of well interval velocity data is the generally sparse spatial sampling of this information with respect to the expected velocity variation in the area. In contrast, velocity information derived from seismic common midpoint (CMP) gathers is generally highly sampled spatially. This advantage is partially offset by the lower resolution of the results. Optimization of the available velocity information is hopefully achieved through the integration of the two separate interval velocity data types, and utilization of this information in the production of spatially variant interval velocity fields. Where problems were encountered in tying the well and seismic data, the optimal approach was considered to be an interpretive "best fit" solution. In these cases consideration of geology, structuring of the overburden, and the calculated interval velocities of both the over- and under-lying section became particularly important.

A variety of software products, located both on mainframe and personal computer hardware platforms, were utilized to produce the final depth maps (Figure 7). Basic mapping work was completed on a mainframe based in-house proprietary mapping system. Well interval velocity analysis from checkshot data was performed on personal computer based in-house software.

The interval velocity for the Recent to Pliocene section was generated as a spatially variable, but vertically constant, interval velocity map.

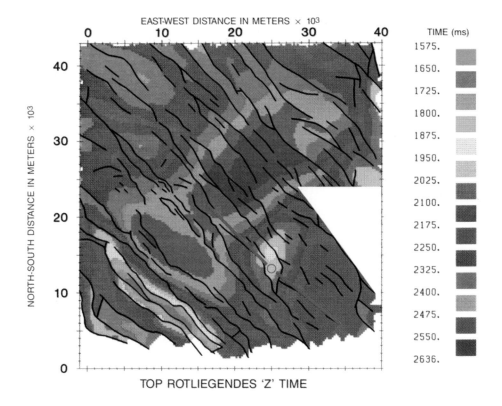

Fig. 5. Stacked (unmigrated) time map (Base Zechstein/Top Rotliegendes) of the interpreted seismic data showing the dominant northwest trending pattern of faulting at this level. The time structure is dominated by the salt pullup effect of the overlying Zechstein (compare with Figure 3). The solid red line is the location of the structural cross section shown in Figure 2. The red circle is the location of the 49/2-3 well. Nine other wells are located within the study area (not shown).

Interval velocities for the Tertiary, Cretaceous Chalk, and Triassic intervals were based on a 3-D, spatially variant, velocity function. The general form can be described as follows:

Given the three interval velocity functions:

$$VI0_{xyz} = V0_{xy} \qquad \text{(2-D function)}$$

$$VI1_{xyz} = V1_{xy} + (K1*Z) \qquad \text{(instantaneous function of depth)}$$

$$VI2_{xyz} = V2 \qquad \text{(constant)}$$

where:

$VI0$, $VI1$, $VI2$ = interval velocities,

$V0$, $V1$, $V2$ = initial velocities,

$K1$ = velocity gradient, and

Z = depth.

Note: subscripts denote the spatial variation of the parameter.

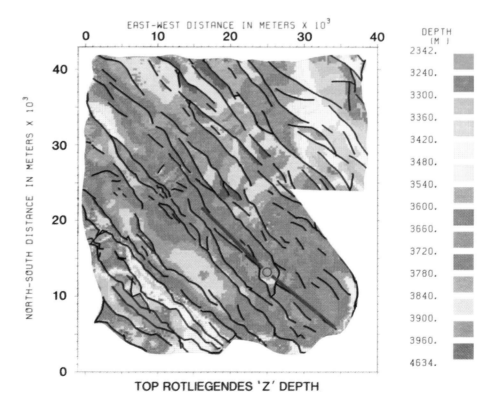

Fig. 6. 3-D migrated depth map (Base Zechstein/Top Rotliegendes) of the interpreted seismic data. The dominant structural pattern is seen to be northwest trending, coincident with the interpreted fault pattern. Note the difference in trend of the primary structure of interest, compared to the time map (Figure 5). The solid red line is the location of the structural cross section shown in Figure 2. The red circle is the location of the 49/2-3 well. Nine other wells are located within the study area (not shown).

The instantaneous interval velocity at a location (x, y, z) is determined from:

$$VI_{xyz} = VI0_{xyz} \text{ when } VI0_{xyz} > VI1_{xyz},$$

$$VI_{xyz} = VI1_{xyz} \text{ when } VI1_{xyz} > VI0_{xyz} \text{ and } VI1_{xyz} < VI2_{xyz}, \text{ and}$$

$$VI_{xyz} = VI2_{xyz} \text{ when } VI2_{xyz} < VI1_{xyz}.$$

A 1-D (depth, Z) representation of this function is shown in Figure 8. For this study both the first ($V0_{xy}$) and second ($V1_{xy}$) initial velocities were allowed to vary spatially in a map (i.e., grid) sense; $V2$ and $K1$ were set to a constant. The intention was that $V0_{xy}$ would account for uplift and truncation effects, while $V1_{xy}$ would account for velocity variation due to changes in lithology. The function $V1_{xy}$ is derived from the available well and seismic data from all areas which are considered to be at their maximum depth of burial. The method involves subtracting from the observed interval velocity the product of $K*Z_m$ (where Z_m is the midpoint depth of the interval under analysis) and gridding the result. The function $V0_{xy}$ is derived from the available interval velocity data in areas where major uplift has occurred. The constant $K1$ represents the average velocity (compaction) gradient for the interval. The third velocity ($V2$) is held constant and is used as a maximum (i.e., "critical") velocity for each interval. Although higher order curves may be used to describe the relationship between maximum depth of burial and velocity {e.g., "Faust's Law" $VI = V1*[Z**(1/V2)]$, Faust, 1953; Sheriff, 1976}, it then becomes difficult to compensate for other factors. For this study utilization of a function/method which would allow for full inclusion of the noncompaction velocity factors (e.g., uplift, truncation, and stratigraphic variation) was considered most critical.

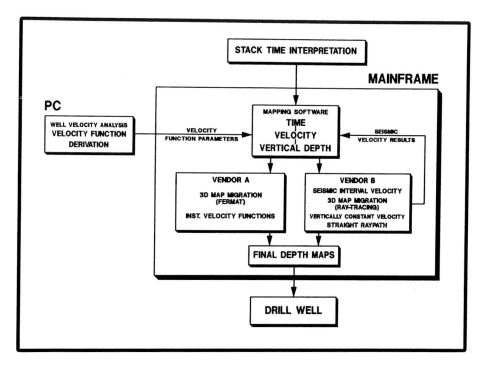

Fig. 7. Pre-drill job flow. A variety of personal computer and mainframe based software products were utilized for this project. All data transfer between software products was done through direct reformatting of the output files. The results from iterative seismic interval velocity analysis and map migration were used in the computation of the final input to the vertical depth conversion and the two 3-D map migration programs.

A further advantage is that this function can be applied as both a true instantaneous velocity function for 3-D map migration and as a midpoint-depth interval velocity for vertical depth conversion in the following manner:

Given that for a vertical raypath the following midpoint-depth interval function velocity is equivalent to the linear instantaneous velocity function previously described

$$VI1_{xy} = V01_{xy} + (K1 * Z_m),$$

where

$$Z_m = \text{midpoint-depth of the interval,}$$

$$Z0_{xy}, Z1_{xy} = \text{depths at the top and base of the interval, and}$$

$$T_{xy} = \text{one-way seismic isochron of the interval.}$$

Since

$$Z_m = (Z0_{xy} + Z1_{xy})/2$$

$$Z1_{xy} = Z0_{xy} + (VI1_{xy} * T_{xy})$$

and substituting

$$Z_m = [(VI1_{xy} * T_{xy}) + (2 * Z0_{xy})]/2$$

$$VI1_{xy} = V01_{xy} + \{K1 * [(VI1_{xy} * T_{xy}) + (2 * Z0_{xy})]/2\}$$

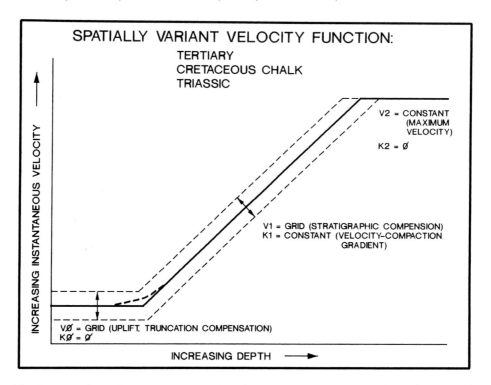

Fig. 8. A one dimensional (z) representation of the 3-D (x, y, z) function utilized in the depth conversion of the Tertiary, Cretaceous, and Triassic intervals. The solid line shows the 1-D depth dependant "guide" function. Parameters $V0$ and $V1$ are input from grids, thereby allowing for 2-D (x, y) spatial variance. The dashed lines show the variation possible in an areal (map) sense. See text for full description.

then rearranging

$$VI1_{xy} = [V01_{xy} + (K1 * Z0_{xy})]/[1 - (K1 * T_{xy}/2)].$$

In this form it is possible to perform an interval based "layer cake" midpoint-depth velocity depth conversion is possible using the two interpreted seismic time horizons, the depth to the upper horizon and the velocity parameters previously derived for the instantaneous velocity functions. For this study, for the "layer cake" vertical depth conversion, the above calculation was performed for $VI1_{xy}$, the second "leg" of the velocity function, and the results were then concatenated with $VI0_{xy}$ (variable) and $VI2$ (constant) using the rules previously described.

For the Zechstein formation the velocities and thicknesses of halite, anhydrite, dolomite, polyhalite, and shale were determined at the well locations and mapped areally. These thicknesses were then reduced to a single compensation thickness map (grid), derived for a single velocity ($V3$, the average velocity of the dominant unit, in this case dolomite). The following lithology ratio based formulas were then utilized:

$$Z(\text{isopach})_{xy} = \{[T_{xy} - (D3_{xy}/V3)] * V(\text{halite})\} + D3_{xy}$$
$$VI_{xy} = Z(\text{isopach})_{xy}/T_{xy}$$

where:

T_{xy} = Zechstein isochron (from the seismic interpretation),
$D3_{xy}$ = non-halite compensation thickness,
$V3$ = non-halite compensation velocity,
$V(\text{halite})$ = halite velocity,
$Z(\text{isopach})_{xy}$ = calculated Zechstein isopach, and
VI_{xy} = calculated Zechstein interval velocity.

Two software products were used for the depth conversion (map migration) portion of the project. One package (Vendor A) is a 3-D map migration program based on the Fermat principle (Sattlegger, 1985) which is capable of using instantaneous velocity functions. For this implementation, the full 3-D instantaneous velocity functions were input to the software. The second package (Vendor B) is a 3-D ray-tracing based algorithm (Hubral, 1977). This software can utilize a spatially variable, but vertically constant, interval velocity map (single valued per ray) and contains a straight raypath assumption. The software can also utilize a velocity gradient function referenced to an overlying surface (i.e., the top of the interval for which the base is to be migrated). Unfortunately, to directly transform an instantaneous velocity function to this sort of velocity gradient function is not possible. In addition, since the top of the Plio-Pleistocene Unconformity (migration layer 2) was relatively flat (compared to the base of the interval), a velocity gradient function for the underlying Tertiary section which utilizes this surface as the reference horizon would not adequately describe the velocity field. The velocity fields input to Vendor B's software were, therefore, generated from the midpoint-depth based interval velocity calculation previously described.

In order to obtain additional interval velocity information, Vendor B's seismic interval velocity analysis software package was utilized. The objective of this software is to provide an overlying structure and dip compensated interval velocity from seismic CMP gathers and the interpreter's 3-D time surfaces. The analysis is performed on an interval-by-interval basis with a depth model for the overlying section having been determined from previous runs of the velocity analysis and map migration software. A set of moveout curves are computed using arbitrary source-receiver ray-tracing (Hubral based) which duplicates the seismic acquisition parameters. The total moveout times are calculated and the interval velocity is derived from the residual moveout required through the interval under analysis. Reduction from seismic interval velocity to the velocity function parameters $V0_{xy}$ and $V1_{xy}$ was done utilizing the methods previously described and the depth migrated (Hubral based algorithm) horizon maps. Through this technique it was possible to utilize the seismic interval velocity information for both the Hubral and Fermat based map migrations and the vertical depth conversion.

An example spectral display available to the interpreter for analysis of the seismic interval velocity results is shown in Figure 9. The analysis shown was performed for the Tertiary section, which is to the base of migration layer three. The optimum (horizon tied) velocity is seen as a pick at 1600 ms. In addition, the velocity inversion at the base Cretaceous chalk (migration layer four) is clearly seen on Figure 9. The velocity gradually increases from 1600–2000 ms and then suddenly decreases at 2150 ms. This variation is the result of the inclusion in the analysis of the underlying Cretaceous chalk with its higher interval velocity, followed by the (lower velocity) Triassic below 2150 ms.

COMPARISON OF 3-D MAP MIGRATION ALGORITHMS

Parallel processing through the two different map migration algorithms and vertical depth conversion allowed for the direct comparison of the methods. The results are summarized in Figure 10. Depth profiles, along the flank of a salt swell, are shown for both the Top Zechstein and Base Zechstein/Top Rotliegendes horizons. Note the difference in vertical scale between the two horizon profiles.

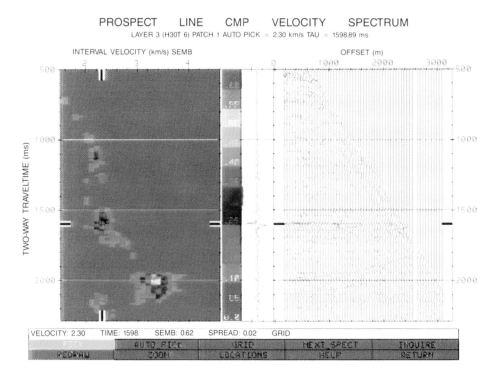

Fig. 9. Example of seismic interval velocity analysis display. On the left is a spectral display of the interval velocity semblance calculation. Moveout curves were derived using a depth/velocity model for the overlying interfaces. The horizontal crossbars show the horizon tied automatic pick for the lower Tertiary section (to Top Cretaceous chalk). Note the increase in velocity with depth and velocity inversion at the base Cretaceous chalk (2150 ms). On the right is the constant velocity moveout corrected CMP gather. In the center is the constant velocity stacked trace.

The differences seen at the Top Zechstein are the primary cause of depth variation at the Base Zechstein/Top Rotliegendes level. The most important result, common to both the Hubral and Fermat based map migration outputs, is that the Base Zechstein/Top Rotliegendes fault trace is moved structurally up dip when compared to its (unmigrated) time and vertical depth conversion position. Since time-domain seismic migration algorithms will generally move reflection events up dip in the time domain, severe positioning error can occur when this time dip direction is not coincident with structural (depth) dip. The time pullup over the profile shown in Figure 10 exceeds 400 ms at the Base Zechstein/Top Rotliegendes level. This time pullup creates a time dip (down to the left) which is opposite to the true structural dip direction. Time-domain seismic migration, therefore, will place the fault plane up time dip, but down structural (depth) dip. The resulting total difference (error) in fault plane positioning between the seismic time-migrated and depth migrated output over this profile exceeds 1.5 km. Note that the dip at the (objective) Top Rotliegendes level does not exceed 3 degrees over the cross section. Further differences between the three depth conversion approaches can be seen at the Base Zechstein/Top Rotliegendes level. The vertical depth conversion, which does not compensate for the migration effect of either the time or velocity fields (since the midpoint-depth velocity calculation is based on the unmigrated time surfaces), produces the shallowest depth at Top Zechstein level and deepest depth (because of the thick Zechstein isochron) at the Base Zechstein/Top Rotliegendes level. Vendor B's (Hubral based) algorithm migrates the time data but uses what, in this application, is effectively an "unmigrated" velocity field. Again, this results because the input velocity fields were calculated using the midpoint-depth functions, which were based on the (unmigrated) time surfaces. The result is a slightly deeper Top Zechstein profile than the vertical depth conversion and a much shallower (because of reduced Zechstein isochron)

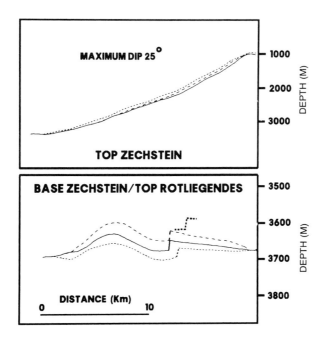

Fig. 10. Comparison, along a salt flank, of the three (Fermat, Hubral, and vertical depth) depth conversion methods. Note the difference in vertical scale between the Top Zechstein and Base Zechstein/Top Rotliegendes profiles. The vertical depth conversion results are shown as a short dashed line; the output from Vendor B's (Hubral-based) map migration is shown as a long dashed line; and the output from Vendor A's (Fermat-based) map migration is shown as a solid line. The heavy dotted line is the position of the fault trace as seen on the time migrated seismic section. See text for full discussion.

Base Zechstein/Top Rotliegendes structural profile. Vendor A's Fermat based depth migration algorithm takes account of both the vertical and lateral variations in the velocity field (through the application of instantaneous velocity functions). This algorithm results in the deepest Top Zechstein and an intermediate Base Zechstein/Top Rotliegendes structural profile. Maximum differences between the three outputs for a salt flank dipping 25 degrees is 100 m. The resultant difference at Base Zechstein/Top Rotliegendes level is 50 meters. Where the dips of the objective and overlying horizons are flat, the difference between the three methods is, of course, exactly zero.

DRILLING RESULTS AND POST-DRILL MIGRATION/MODELING

In 1988 Mobil and Partners drilled the 49/2-3 wildcat well. All mapped horizons were encountered within 15 m of the (Fermat based depth migration) prognosis. As the objective Rotliegendes structure trended oblique to the overlying salt swell, we realized that information derived from a vertical well velocity survey would be of limited use in further delineation of the structure. The critical structural closure (spillpoint) was mapped as occurring beneath the southeastern Zechstein syncline in an area of no well control (Figures 2 through 6). The principle uncertainty was in the derivation of the Cretaceous chalk interval velocities. From the northwestern syncline, where there was thought to be adequate velocity control, to the southeastern syncline, the chalk thickness increased almost two-fold. This (unsampled) stratigraphic variation had the

POST-SURVEY MODEL

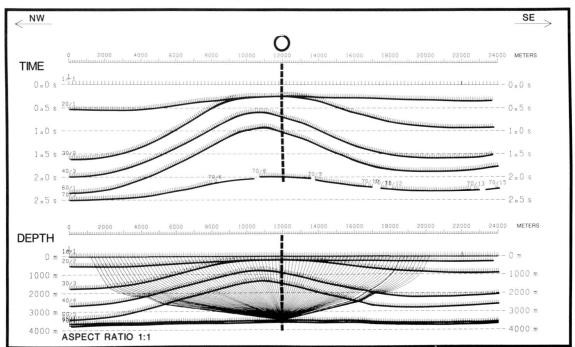

Fig. 11. Stacked (unmigrated) time and 3-D depth migrated display of the walkaway survey "Most Likely" model. The 49/2-3 well location is shown as a black circle. Critical structural closure (spillpoint) of the Rotliegendes surface is at a distance of 17 to 22 km (5 to 10 km offset southeast from the well). Note the raypath curvature through the Cretaceous chalk interval (between horizons 30 and 40). As described in the text, the geophone is located immediately above the Zechstein basal anhydrite/dolomite sequence. The profile of this line is the same as the structural cross section shown in Figure 2.

potential to cause significant error in the calculated interval velocity field. Long offset walkaway survey modeling was undertaken to determine if the first arrival, peg-leg and shear data could be used to derive velocity information from the southeastern syncline. A geophone was located immediately above the basal Zechstein anhydrite/dolomite sequence and a walkaway survey was modeled for source positions 0 to 12 km away from the well location into both the northwestern and southeastern synclines (Figures 2 and 11). An initial assumption was that with the geophone placed above the basal anhydrite/dolomite sequence, the overlying Zechstein (predominantly halite) velocity could be considered to be essentially constant. Any differences between the field results and the modeling experiment could, therefore, be attributed to velocity differences in the post-Zechstein section. Two structural models were tested. The first was the preferred or "Most Likely" case. For the second model the Cretaceous chalk velocity in the southeastern syncline was reduced so that structural (four-way dipping) closure at the Top Rotliegendes was lost. This became the "No Closure" model. The modeling experiment revealed differences between the Most Likely or preferred structural model and No Closure structural of 30–50 ms in the first arrival data. These differences were thought to be measurable in the field and sufficient to justify the long offset walkaway survey.

The survey was shot using a 4 × 160 cu. in. airgun array source, and a three component geophone in the well. Total time for the walkaway portion of the survey was five hours and this was shot in conjunction with a conventional VSP. The data were processed using source-signature deconvolution, minimum phase

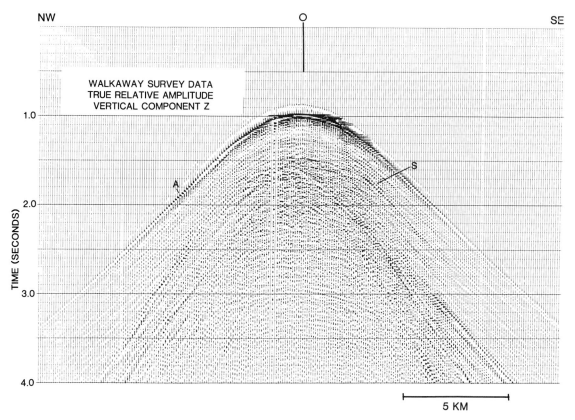

Fig. 12. True relative amplitude display of vertical component (Z) geophone. As described in the text, these data have been processed and scaled to preserve the true relative amplitudes of all arrivals. As a result, the near-offset direct arrival is severely over-gained. Note the high energy shear wave (S), and the first arrival amplitude bloom (A) at 6 km northwest offset. The well location is shown as a black circle. The survey orientation is the same as in all preceding vertical displays.

filtering (2/5 Hz–80/120 Hz), predictive deconvolution (400 ms operator, 50 ms gap, 1000 ms window), followed by zero phase filtering (2/5 Hz–80/120 Hz). The vertical and two-horizontal component geophone data were then displayed as True Relative Amplitude sections. The vertical component (Z) geophone data are shown in Figure 12. Three component walkaway survey processing (as described in Noble et al., 1988) was not attempted since the single shot level was located to provide maximum direct arrival, long offset coverage. This location was incompatible with successful imaging of the primary reflector of interest (Base Zechstein/Top Rotliegendes).

As a result of the presurvey modeling it was possible to immediately compare the field results with the existing structural models. A difference plot between the field results and the Most Likely and No Closure structural cases is shown in Figure 13. Note that the difference between the two structural models exceeds 50 ms at 6 km southeast offset from the well. The Most Likely model more closely matched the field data than the No Closure case (difference less than 10 ms from 0 to 6 km southeast offset), therefore immediately confirming structural closure to the southeast. Severe mismatch was present however in the northeast syncline portion of the walkaway survey. This was in an area where well control was available.

The post-drill job flow is shown in Figure 14. The well VSP velocity information, walkaway survey velocity information, and formation tops were incorporated into the velocity function calculations and time maps. This (revised) information was then input into the two map migration algorithms, followed by walkaway survey modeling over the (revised) depth migrated surfaces. The model output was then compared with the field results and the entire process iterated until the observed differences were minimized. Again, the parallel processing allowed for direct comparison of the two vendor software packages. The significant differences between the two modeling algorithms are as follows. Vendor A's well seismic modeling software is a 2-D algorithm based on ray-tracing but using curved raypaths and instantaneous velocity functions. Vendor B's modeling software is a 3-D algorithm based on ray-tracing using straight rays and spatially variable, but vertically constant (single valued per ray), interval velocity fields.

The post-survey modeling procedure involved generation of amplitude events from 23 separate arrivals. These included the direct arrival, peg-leg multiples, converted S-waves, S-wave multiples, and reflection events. Of the 23 modeled events five distinct amplitude events were obtained (Figure 15). These included S-waves generated from the Base Cretaceous chalk (4TS6Q) and Top Zechstein salt (5TS6Q), the direct arrival and, immediately following the direct arrival, two reflection events from the top and base of the basal anhydrite/dolomite sequence. The modeling results very closely match the actual field data. The direct arrivals, the two reflection arrivals, and the S-wave energy are clearly seen on the vertical component (Z) geophone True Relative Amplitude display (Figure 12). Data from the two horizontal component geophones (not shown) confirm the interpretation of high amplitude S-wave energy being present in the field data. After iterative map migration and modeling the final minimum difference result was achieved (Figure 16).

The residual error was less than 10 ms at a distance of up to 6 km away from the well bore. The southeastern syncline residual error was less than 20 ms out to a distance of 10 km from the well bore. This is equivalent to a maximum depth error at Top Rotliegendes, attributable to velocity error for that section investigated by the walkaway survey, of approximately 15 m. The deviation on the difference plot between the field and modeling results (Figure 16) was not significantly reduced in the northwestern syncline. The extensive post-drill well seismic modeling indicated that this apparent anomaly was due to the generation, at a distance of 6 km

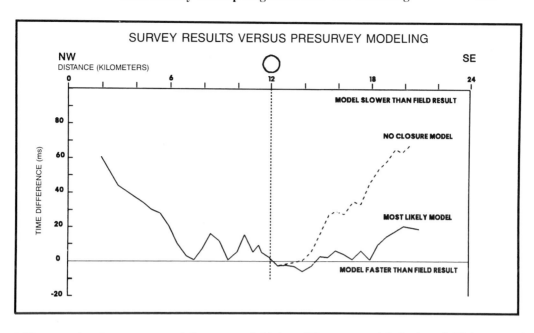

Fig. 13. Difference plot; Pre-survey Modeling versus field data difference (modeled minus field) between the first arrival as recorded on the field data and the two models. The well location is shown as a black circle. The orientation is the same as in all preceding vertical displays.

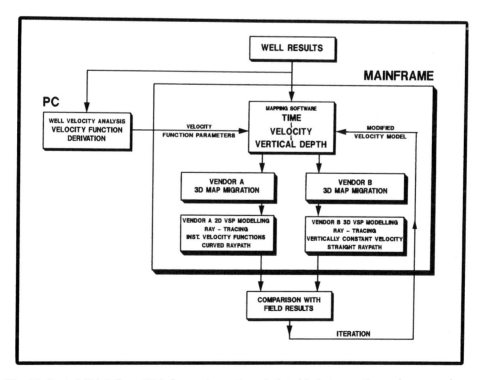

Fig. 14. Post-drill job flow. This figure shows the relationship between the various mapping, seismic, and well velocity analysis, map migration and modeling software utilized in the post-drill job flow. The migration and modeling routines were iterated until the difference with the field walkaway survey data was minimized. See text for details.

northwest of the well bore, of head wave refractions from the top and base of the basal Zechstein anhydrite/dolomite sequence. These arrivals overtake the direct arrival and are recorded as the first arrival on the field data. The modeling results are confirmed by phase and amplitude changes observed on the vertical component (Z) geophone data (Figure 12). This interpretation is also confirmed by data from the two horizontal component geophones (not shown). A difference plot between the modeling results of the two vendor software products is also shown in Figure 16. The near offset difference is due to the 2-D modeling assumption of Vendor A's algorithm. Since there is a moderate off-line dip at zero offset, Vendor A's results are in error at this location. A linear systematic error or difference between the two software packages also is seen to the southeast of the well bore.

CURVED-RAY VERSUS STRAIGHT-RAY MODELING RESULTS

In an attempt to resolve differences observed between the two vendor software products, comparative curved-ray versus straight-ray modeling was done using Vendor A's algorithm. A comparison of the two model outputs is shown in Figure 17. Significant variation in critical angle between the curved-ray and straight-ray approaches was observed. Major differences also can be seen in raypath refraction at the Base Cretaceous chalk. This raypath refraction is primarily the result

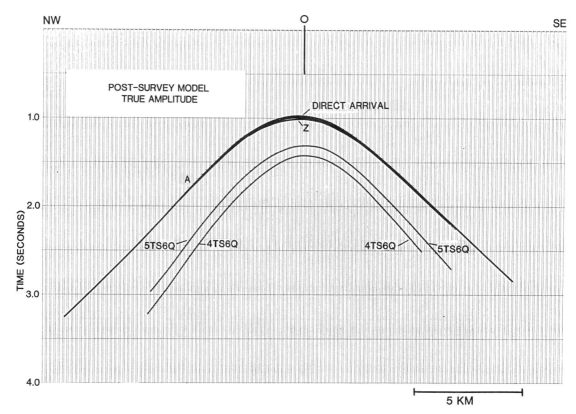

Fig. 15. Five principle amplitude events were recovered from the modeling of 23 possible events. As discussed in the text, significant amplitudes were computed for the direct arrival, top and base of the Zechstein basal anhydrite/dolomite (Z, shown together), and transmitted S-wave energy from the base Cretaceous chalk (4TS6Q) and Top Zechstein (5TS6Q). Note the convergence of direct arrival and Zechstein basal anhydrite/dolomite reflection events at 6 km northwest offset (A). Compare with Figure 12. The well location is shown as a black circle. The orientation is the same as in all preceding vertical displays.

of the difference between the gross velocity of an interval and the instantaneous velocity at an interface boundary. In this case the actual velocity at the base Cretaceous chalk is much greater (6700 m/s) than the gross interval velocity (4800 m/s). Use of an instantaneous interval velocity model also results in the generation of a much shallower velocity inversion effect at the base of the formation. The curved-ray, instantaneous-velocity method much more closely approximates the "true" geologic velocity field for this interval (as determined from the available well wireline data).

A difference plot between curved-ray and straight-ray results is shown in Figure 18. In the southeast syncline the deviation almost exactly matches the Vendor A versus Vendor B results difference plot (see Figure 16), indicating that the observed variation in the modeling results is due almost entirely to the curved-ray versus straight-ray approaches. Theoretically, one could approximate the curved-ray, instantaneous-velocity method using a straight ray, vertically constant, velocity algorithm through fine subdivision of the interval of interest. In practice, however, this is extremely cumbersome and causes other problems, chiefly the build-up of ray-tracing "noise".

The difference in software algorithm capability can be very important, not only in well seismic and Normal Incidence modeling but also in surface seismic Arbitrary Source-Receiver models, where simulation of the true impedance contrasts at interface boundaries is often critical.

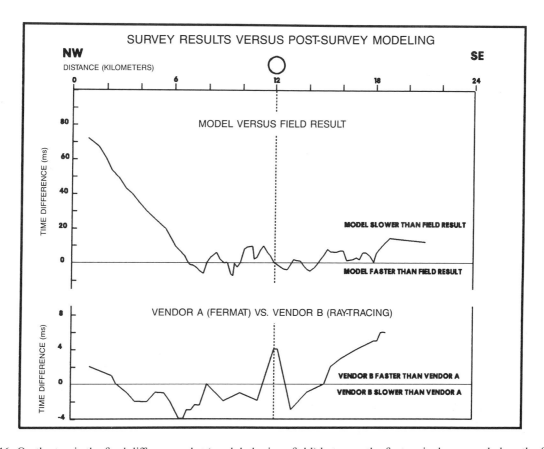

Fig. 16. On the top is the final difference plot (modeled minus field) between the first arrival as recorded on the field data and the final depth model. As discussed in the text, the difference at 6 to 12 km northwest offset is due to first arrival of the critical refraction from the top of the Zechstein basal anhydrite/dolomite. This precedes the direct arrival and cannot be modeled using the available software products. On the bottom is a difference plot between the results of the two vendor software products. See text for full discussion. The well location is shown as a black circle. The orientation is the same as in all preceding vertical displays.

VELOCITY ANISOTROPY

One of the problems considered during the course of this study was the importance and effect of velocity anisotropy on the migration/modeling process. Any significant amount of velocity anisotropy would result in the seismic derived interval velocity being in error. The error would occur because seismic interval velocity derived from the moveout velocity is considered to be a measurement of the horizontal component of the velocity field (Vander Stoep, 1966; Levin, 1978). Also at far offsets the modeling routines, because they only allow for isotropic velocities, could produce results that were inconsistent with the field results due to the anisotropic effect.

Banik (1984) compared stacking velocity to well-log rms velocities in 21 data sets in the North Sea. A correlation was obtained between the presence of shales and the calculated velocity anisotropy. Anisotropy effects of up to 20 percent were computed for the Tertiary and Jurassic sections (the Triassic was not evaluated). The Cretaceous chalk was considered to be isotropic.

Although differences in the seismic-derived and well-derived interval velocities were observed in the seismic velocity analysis phase of this project, they did not appear to correlate with sand/shale ratio variations calculated from the well data. Large seismic interval velocity scatter was observed which could be directly correlated with either structural dip, ray-tracing "noise" in the moveout curve calculation, or errors in the velocity analysis of overlying horizons.

For this study, the interval velocity functions computed from the well and seismic data were derived so that the output maps tied the well vertical seismic (checkshot and/or VSP) results exactly in depth. Therefore, the interval velocities

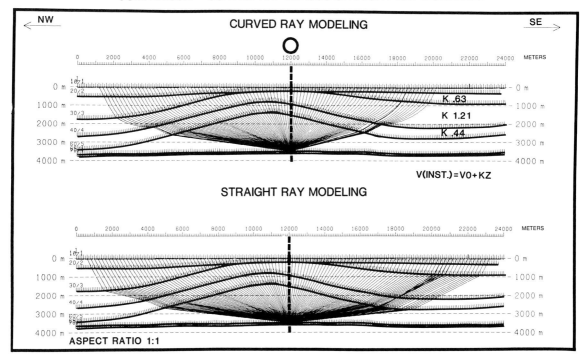

Fig. 17. Walkaway survey model displays using straight-ray and curved-ray assumptions. The 49/2-3 well location is shown as a black circle. Note the differences in (1) raypath through the Cretaceous chalk interval (horizons 30 and 40), (2) critical arrival angle, and (3) the effect of the velocity inversion at the base of the Cretaceous chalk (horizon 40). The velocity gradient factor K is shown for each interval. The geophone is located immediately above the Zechstein basal anhydrite/dolomite sequence. The profile of this line is the same as the structural cross section shown in Figure 2.

used in the migration/modeling process were considered to be the vertical component of the compressional velocity field. If significant (15–20 percent) velocity anisotropy existed in the Tertiary and Triassic section, the walkaway survey far offset first arrivals would be "faster" than the modeled arrivals (assuming a higher horizontal component velocity) by 55 to 75 ms. Although raypath angles of up to 60 degrees from horizontal were calculated over the model, no differences of this magnitude were observed. Since the field direct arrival data appeared to tie the "known" (well tied) vertical velocity field out to a distance of 6 km offset into the northwest syncline, anisotropy effects on the compressional velocity field were considered to be minimal in this area.

CONCLUSIONS

The integration of (1) well and seismic interval velocity data and (2) 3-D map migration and well seismic modeling, allowed for the resolution of the structural configuration at the objective horizon to within an estimated error of approximately 15 m. Several strengths and weaknesses were observed in the various methodologies utilized for depth conversion in the area. The primary job of the seismic interpreter is to identify that software product or methodology which yields the most reliable result for a given problem.

Differences were noted between 3-D map migration results for relatively low dips at the objective interface when the structural dip of the overlying section exceeded 15 degrees. These were primarily due to the application of gross midpoint-depth derived interval velocity versus instantaneous interval velocity functions in the depth migration process. Comparison of the Hubral and Fermat based algorithms revealed that the variations observed were generally insignificant when the dip of the overlying section was less than 25 degrees. When the dip of the overlying section exceeded 25 degrees significant differences do begin to appear. However, a problem exists in that at higher dips the interpretation itself

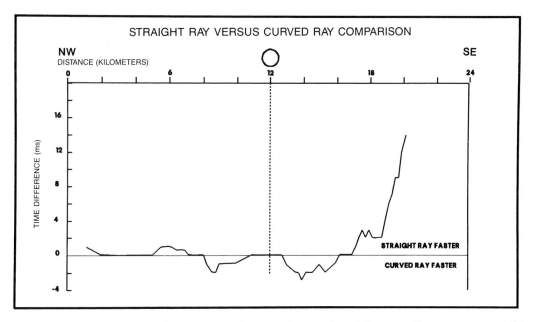

Fig. 18. Difference in direct arrival time between the curved-ray and straight-ray walkaway survey models. Note the gradual increase with offset in the southeast direction (location 14 to 20 km on the model) and similarity to the difference observed between the results of the two vendor software products (Figure 16). The well location is shown as a black circle. The orientation is the same as in all preceding vertical displays. See text for details.

becomes fairly subjective and subject to error. Therefore, for high-dip geometries comparison of algorithms is best done on synthetic, rather than real, datasets.

Although the question of stratigraphic anisotropy was considered in the evaluation of the long offset modeling, no evidence was found in the field data to support a significant effect.

Finally, a new generation of migration and modeling software has emerged which has many applications in exploration and production geophysics, can have direct and immediate impact on the drilling program, and yields results to within an accuracy not previously attainable.

ACKNOWLEDGMENTS

I thank Mobil North Sea Ltd. and partners Coalite Group Plc, Fina Exploration Limited, and Sovereign Oil & Gas Plc for permission to publish this paper. In addition, the following companies are acknowledged for their permission to publish interpreted exploration data over held acreage:

City Oil Exploration Limited	Clyde Balmoral
Esso Exploration & Production U.K. Ltd.	Enterprise Oil Plc
Fina Exploration Ltd.	Lasmo North Sea Plc
Nedlloyd Energy Ltd.	Occidental Petroleum Co.
Sante Fe Minerals U.K. Inc.	Shell U.K. Explor. & Prod.
Texas Eastern (U.K.) Ltd.	Thomson International
Total Oil Marine Plc	Unocal U.K. Ltd.

I also acknowledge Sattlegger Ingenieurburo Fur Angewandte Geophysik, Sierra Geophysics Inc., and Seismograph Service Ltd. for their substantial contribution of both software and technical expertise to this project.

REFERENCES

Banik, N. C., 1984, Velocity anisotropy of shales and depth estimation in the North Sea basin: Geophysics, **49**, 1411–1419.
Bulat, J., Stoker, S. J., 1987, Uplift determination from interval velocity studies, U.K. southern North Sea, in Brooks, J., Glennie, K., Eds., Petroleum geology of North West Europe: Graham and Trotman, 293–305.
Carter, M. D., 1987, Velocity interpretation: Southern North Sea Gas Basin Case Study: 57th Ann. Internat. Mtg., Soc. Expl. Geophys., Expanded Abstracts, 427–428.
Dix, C. H., 1955, Seismic velocities from surface measurements: Geophysics, **20**, 68–86.
Faust, L. Y., 1953, A velocity function including lithologic variation: Geophysics, **18**, 271–297.
Hubral, P., 1977, Time migration—Some ray theoretical aspects: Geophysical Prospecting, **25**, 738–745.
Levin, F. K., 1978, The reflection, refraction, and diffraction of waves in media with elliptical velocity dependence: Geophysics, **43**, 528.
Nobel, M. D., Lambert, R. A., Ahmed, H., and Lyons, J., 1988, The application of three-component VSP data on the interpretation of the Vulcan Gas Field and its impact on field development: First Break, **6**, no. 5, 131–149.
Sattlegger, J., 1985, Map migration and modeling algorithm: 55th Ann. Internat. Mtg., Soc. Expl. Geophys., Expanded Abstracts, 553–554.
Sheriff, R. E., 1976, Encyclopedic dictionary of exploration geophysics: Soc. Expl. Geophys.
Vander Stoep, D. M., 1966, Velocity anisotropy measurements in wells: Geophysics, **31**, 900–916.

INDEX